THE PARADOX OF PROGRESS

THE PARADOX OF
PROGRESS

BOOK II:
THE ROSES AND THORNS OF
ARTIFICIAL INTELLIGENCE

MICHAEL M. KARCH, MD

HOUNDSTOOTH
PRESS

THE PARADOX OF PROGRESS
Book 2: The Roses and Thorns of Artificial Intelligence

FIRST EDITION

ISBN   978-1-5445-4929-3 *Hardcover*
        978-1-5445-4928-6 *Paperback*
        978-1-5445-4927-9 *Ebook*

Dedicated to my mother, Lillian, who always saw humanity through rose-colored glasses, and to my father, Peter, who brought awareness to our thorns.

Technology, life, and truth likely lie somewhere in between.

CONTENTS

"We must work passionately and indefatigably to bridge the gulf between our scientific progress and our moral progress."

—DR. MARTIN LUTHER KING JR.

"The presence of challenges alongside blessings."

—PAUL THE APOSTLE, 2 CORINTHIANS 12:7 (KOINE GREEK), AD 56

"Pas de rose sans epine. (No rose without a thorn.)"

—OLD FRENCH FOLK SAYING

INTRODUCTION

Thanksgiving Week 2020
Von's Grocery Store. Mammoth Lakes, California

Armed with a long grocery list from my wife, I gave my middle-aged brain a workout, reminiscent of a Sudoku puzzle.

I was determined to memorize the entire Thanksgiving grocery list, channeling the little girl from a 1972 *Sesame Street* cartoon "I Can Remember" who famously chanted, "A loaf of bread, a container of milk, and a stick of butter."[1] She repeated her mantra all the way to the store, only to stand befuddled amid the aisles once she got there.

Overconfident in my own skills of recall, I soon found myself staring blankly at the grocery shelves similar to the cartoon from so many years ago.

"Dad," my youngest daughter, Olivia, chimed in, her tone laced with that unique teenage impatience. "Why didn't you

---

1   *Sesame Street*, Season 4, Episode 8, "I Can Remember," art by Jim Simon, directed by Sam Gibbon, aired November 15, 1972, on PBS.

just dictate the list into your phone?" I grudgingly gave in, the two of us rattling off items from memory. Yet somehow, between the two of us, we still missed a few things.

Olivia was right. The human memory for long lists is just not that good, at least not mine.

When we finally hit the checkout, I was greeted by long, snaking queues of masked-up tourists. COVID-19 was still in full swing, and the guy a few people over, who assured everyone that his watery eyes were "just allergies," had me second-guessing my place in line. That's when I decided to try the self-checkout for the first time—surely it'd be quicker and less congested.

Having paid far too much in my lifetime to insure my hands as a surgeon, I cautiously passed one hand under the infrared light to see if they were actually worth anything. Nope, it seemed that without a barcode in the modern-day world, you don't count for much. Scanning my grocery items was quite easy—and oddly satisfying, each beep and blurp adding to a little self-checkout concerto.

Hey, this automatic thing was kind of fun.

Then came the bag dilemma. The machine inquired how many I needed. A devilish thought crossed my mind: *What if I entered o and sneakily snagged a bag? What if I "forgot" to scan one avocado or one clump of juicy red grapes? Would the grocery gods smite me on the spot? Would alarms blare, proclaiming my petty thievery to the world?* I glanced around, half expecting to see the armed security guards closing in.

Chuckling, I keyed in five bags, abandoning my brief career in petty crime.

As I bagged everything, my collar up and my eyes half turned away from the hidden camera, I couldn't help but wonder—do guilty thoughts count if you don't actually do the deed?

Are thoughts of crime a real thing?

The self-checkout process proved to be a swift affair, significantly quicker than the traditional checkout line. As I completed the transaction, the machine uttered a cold robotic, "Thank you." Efficiency was clearly on my side as I loaded my groceries into the car, far ahead of those still queued in the regular line.

Yet, despite the speed, there was a noticeable lack of warmth in the experience. I found myself missing the familiar, friendly exchange with Bonnie, the checkout lady who had been a part of my grocery routine for the last twenty years. Her genuine smile and heartfelt, "Happy Thanksgiving" always added a human touch to the chore, something that the efficient but impersonal machine could never replicate.

A UNIQUE INTERSECTION

"The great paradox of our time: Everything is both better and worse than it ever has been before."

—ROB WIJNBERG

My little grocery store escapade—filled with the struggle of recalling items, navigating tech-savvy teens, and having a mild existential crisis over self-checkout—got me thinking about artificial intelligence and how deeply it's embedded in our lives. We're at a unique intersection, living in a world where AI is seamlessly woven into the fabric of daily life. Have we really reached a point where everything is better *and* worse than ever before? And where does this road with AI lead?

Is this true progress?

Maybe you're excited about the future of AI and are enjoying a more streamlined life. Unlike me, maybe you routinely dictate your grocery list into your phone, never missing an item. Maybe you instruct Alexa to play music with a simple command or to dim your lights before watching a movie with your family. Maybe you pay for Starbucks with your smartwatch, no longer carrying a credit card, let alone cash in your wallet. If so, you might view AI as wonderful progress.

Or, maybe you're in the opposite camp: skeptical of what the future of AI may bring. Maybe you feel as though you are being left behind at the speed in which AI is being adopted and are simply ignoring its infiltration. Maybe as AI systems become more autonomous, you fear we may lose control over its actions. Maybe you're scared it'll take our jobs, or that our privacy could get compromised and fall into the wrong hands.

Neither view is entirely right or wrong. AI sits squarely in a paradox—it's both progress and a potential peril. That's why this book is titled *Paradox of Progress*. AI could unlock remarkable potential and solve problems we once thought insurmountable, or it could shake the very structure of society.

Regardless of which camp you're in, AI is here and it is here to stay. As this technology evolves, we have to ask ourselves serious questions. What parts of human interaction are we willing to trade for speed and convenience? Are we sacrificing something precious as we automate more of our lives?

Progress, after all, doesn't come without costs.

While progress signifies advancement and improvement, the paradox of progress dictates that forward movement is never without its downsides and unintended consequences. Progress is not an absolute good; it is a complex, multifaceted process that often brings new challenges even as it solves old ones.

Just as roses are beautiful, they also have thorns.

AI is no different, and in the chapters ahead, we'll dive into the beauty and complexity of this paradox, exploring how it shapes, complicates, and challenges our world.

INSIDE THESE PAGES

AI holds incredible promise for tackling some of humanity's biggest challenges—from addressing systemic racism and climate change to managing overpopulation and reducing global conflict. I call these the roses. At the same time, AI also brings tough questions and risks, like job displacement, ethical dilemmas, and potential for misuse. I call these the thorns.

This book dives into this dual nature—both humanity's and AI's—showing our capacity for good and bad, creation and destruction, compassion and cruelty, generosity and greed. Progress is never one-sided; every advancement carries its own set of challenges.

Before exploring the roses and thorns of AI, the first few chapters will lay the foundation, giving you the necessary context to better understand the paradoxes ahead.

In Part One, we will cover:

- How past technological revolutions led to unintended consequences—and what they teach us about the future of AI
- How AI is a reflection of us—our strengths and flaws—and what that means for its development
- The crucial role of trust in AI—how data bias, machine errors, and transparency will determine its reliability and future
- How the global race for AI dominance is shaping technology, power, and the future of digital warfare.

In Part Two, we'll cover a carefully selected array of paradoxes associated with the progress of AI, as I perceive them. They are as follows:

- The journey and challenges of self-driving cars

- The environmental and social impact of mining materials for rechargeable batteries
- Balancing AI's high energy use with its ability to drive sustainability
- The effect of automation on jobs and the changing role of humans in an AI-powered world
- How AI can either widen or help close global inequalities
- The rise of deepfake technology and its consequences for trust and authenticity
- The use of AI in military contexts, with ethical implications and the risk of unintended harm

Of course, these are just *some* of the issues—AI's influence stretches far beyond what we can cover in a single book. The goal here isn't to provide definitive solutions to these complex challenges. Instead, it's to spark awareness and encourage you to think critically about the path we're on with AI and the tough decisions we'll face along the way.

WHY I WROTE THIS BOOK

Artificial intelligence embodies the paradox of progress like nothing we've seen before. The familiar tech mantra "Move fast and break things" might have flown during the internet boom, but AI is too impactful to take that approach. It demands something deeper and more thoughtful.

I don't come to AI as a tech industry insider or someone with a financial stake in its development. I'm just an outsider and end user, seeing AI increasingly woven into both my professional and personal life. My journey with AI began through my work with surgical navigational instruments, particularly in orthopedic trauma. Together with two other surgeons, we identified a significant issue in the operating room and sought to resolve it. Using computer navigation, we coinvented a drill system that enabled more precise drilling and measurement techniques during surgery. This effort resulted in multiple patents in medical technology, including the SMARTdrill™—a handheld navigational power drill approved by the FDA for orthopedic trauma surgery. This device uses computer navigation to provide real-time data, ensuring accuracy in drilling, which is critical for successful surgical outcomes.

Following the success of SMARTdrill, I quickly adopted computer navigation for the precise placement of hip replacement components in its early stages. Seeing the clinical benefits for my patients, I integrated the DePuy VELYS™ Robotic-Assisted Solution for knee replacement surgeries. This technology allows for precise removal of damaged bone and accurate joint alignment, leading to better-balanced joints, shorter recovery times, and reduced pain for patients.

Building on VELYS's success, I continued addressing various challenges, earning additional patents in medical technology. I also co-founded SteriTools US and UK Inc., which developed the Steri-BasinGo (Tactical), an intraoperative sterilization device designed to reduce surgical site infections in both military and civilian settings. These early advancements in surgical technology paved the way for my deeper exploration of advanced technology and artificial intelligence.

While computer navigation and robotics are driven by human decisions, artificial intelligence introduces a new frontier where the technology itself can make decisions. This concept intrigued me and led to a two-year deep dive into AI, followed by advanced study at the Harvard Business School in data analytics and AI, as well as executive-level courses at MIT in machine learning. Although this exploration raised more questions than answers, one truth became evident: The general public must be involved in the conversation to fully grasp how AI will shape our lives and the future.

That's why I wrote this book.

Progress isn't just about moving forward; it's about pausing to weigh the benefits and costs, aiming to enhance positive outcomes while minimizing the downsides. This requires a nuanced perspective, one that respects the interconnectedness of society's advances and recognizes the unintended consequences that follow.

This book calls on developers, academics, policymakers, and the public to join in a meaningful dialogue about the paradoxes AI brings. It stresses the importance of embedding ethics into the core of tech innovation so that AI enhances our lives in ways that respect and reflect our best human values.

I invite you to join this essential conversation about AI's impact on society. Rather than take sides, I hope to spark discussions within the global community, exploring both the roses and thorns of AI. This book doesn't argue for or against AI, nor does it claim to have all the answers to the complex questions surrounding this transformative technology.

My aim, instead, is to highlight AI's paradox of progress and how forward movement often brings new problems even as it solves old ones. AI certainly won't solve all of our problems and create a utopian paradise, but it also won't single-handedly cause the destruction of mankind, either.

AI isn't inherently good or bad; it's what *we* make of it.

As such, every step forward must be accompanied by reflection and responsibility. As we harness the power of AI, we must do so with an awareness of its dual potential: to elevate humanity or to deepen existing divides. By acknowledging and addressing the unintended consequences, we can steer progress toward a net positive impact.

Ultimately, embracing the paradox of progress means accepting that our advancements will never be perfect or complete. It means striving for continuous improvement, learning from our mistakes, and being willing to adapt our approaches. It calls for a balanced perspective that recognizes both the promise and the peril of progress, ensuring that our pursuit of a better future is guided by wisdom, inclusivity, and ethical consideration.

To fully grasp where we're headed with AI, we first need to step back and look at the bigger picture. Before diving into AI's paradoxes, Part One will lay the foundation—examining past technological revolutions, the ways AI mirrors both our strengths and flaws, the crucial role of trust, and the global race shaping its future. Understanding this context will help us navigate the promises and challenges AI presents in Part Two.

Let's get started.

PART ONE

LAYING THE FOUNDATION

CHAPTER 1

THE PARADOX OF PROGRESS: UNINTENDED CONSEQUENCES IN HISTORY

"What is past is prologue."

—WILLIAM SHAKESPEARE, *THE TEMPEST* (1611)

August 1835. 94°F
London, England

The smog-laden air and the relentless clatter of machinery formed the backdrop to the life of Eliza, a twelve-year-old orphaned child laborer in a textile factory. Born into the pastoral tranquility of a farm, her life had taken a grim turn once she had moved to the city. The green fields and open skies of her childhood were now replaced by the oppressive walls of the factory and the cramped quarters of urban slums.

Eliza's days were a monotonous blur of endless labor. She worked for many hours under the watchful eyes of factory overseers, her small hands deftly maneuvering the looms and spindles. The factory was a cacophony of noise and danger, where a moment's inattention could result in injury. The air was thick with cotton dust, making each breath a struggle.

Her living conditions were no less harsh. The urban slums were a maze of narrow, dirty streets and overcrowded tenements. Sanitation was a foreign concept here; the water was as murky as the future it reflected. Eliza and her fellow laborers squeezed into these cramped spaces, sharing their misery and meager meals, which did little to fend off the gnawing hunger.

Nutrition was poor, a far cry from the fresh produce of her family's farm. Her father, once a proud and strong farmer, died last year in an industrial accident, a common fate for many in these new and perilous workplaces. This past winter, her mother had succumbed to cholera, a disease that thrived in the overcrowded and unsanitary conditions of the city. Eliza's life had shifted dramatically—from the nurturing embrace of nature to the harsh, unyielding grip of industrialization.

Two years prior, when her parents informed their children they were selling the farm and moving to the city in the hopes of an "easier life," I doubt they foresaw any of this.

If only they had thought ahead.

HISTORICAL LESSONS

The Industrial Revolution posed a formidable "social question" to society, demanding new ideas on how to manage large groups of people in cramped areas. The stark contrast between visible poverty and growing materialistic wealth created palpable social tensions. On one hand, we have Eliza, embodying the struggles and hardships of the poorest in society, and on the other, the factory owners and the new industrialists, basking in unprecedented wealth.

These tensions sometimes erupted violently, giving birth to radical philosophical ideas such as socialism, Marxism, communism, and anarchism. Eliza's story, a microcosm of the broader societal upheaval, mirrored the drastic shift from a simpler, agrarian lifestyle to the complexities and challenges of the industrial age—the paradox of progress.

Using history as a learning tool, we can draw parallels between the current AI revolution and the Industrial Revolution of the late eighteenth century. Both eras represent significant technological shifts that dramatically altered society and the economy. The Industrial Revolution's rapid industrialization, marked by an "all in" mentality, brought remarkable progress but paradoxically overlooked crucial ethical considerations and potential negative consequences—the roses and thorns.

Looking at the roses, the rise of factories mechanized tasks, transforming economies and drawing people to rapidly expanding cities. New machines during this time significantly boosted production rates and efficiency, which led to the mass production of goods, making products more affordable and accessible. The rise of factories also created job opportunities for people, allowing many to earn a living and improve their standard of living. The era also spurred numerous technological innovations, such as the steam engine and telegraph, which obviously revolutionized transportation and communication.

However, this progress came at a high human and environmental cost. Looking at the thorns, workers, including children, faced long hours in dangerous conditions for minimal wages, while the explosive urban growth led to overcrowded, unsanitary living conditions and rampant disease. The social fabric was strained as traditional ways of life were disrupted, and economic inequalities deepened.

These unintended consequences led to political polarization and eventually necessitated evolutionary reforms, such as labor laws, urban planning, and environmental regulations. These changes, albeit vital, were implemented only after and in response to the harms already inflicted, underlining the critical need for foresight and ethical consideration amid transformative changes.

A MORE RECENT EXAMPLE

Okay, you may be thinking, *but that example was almost two hundred years ago.*

Fair enough.

Understanding the impact of historical events from two hundred years ago can be challenging for many. Let's examine a more recent example of how technology's unintended consequences are influencing our lives today.

In a *New York Times* article dated December 11, 2023, reporter David Leonhardt delves into the troubling rise in traffic deaths, a critical issue that underscores the unintended consequences of technological advancements, especially in the realm of vehicular safety and smartphone usage. He traces a historical arc to the 1920s, a period marked by a stark absence of laws, guidelines, and public education on car safety, which contributed to a significant toll on human life from vehicle crashes.[2]

---

2   David Leonhardt, "The Rise in U.S. Traffic Deaths: What's Behind America's Unique Problem with Vehicle Crashes?" *New York Times*, December 11, 2023, https://www.nytimes.com/2023/12/11/briefing/us-traffic-deaths.html.

At the turn of the century in 1900, the rate of roadway human fatalities in the US was notably low, with only 0.05 deaths per 100,000 people, totaling 36 deaths. However, as the use of automobiles increased and roads became more congested, these figures surged dramatically. By 1937, the fatality rate had jumped to 29.36 per 100,000, culminating in an alarming 37,819 deaths, according to the US Motor Vehicles fatalities report.[3] This sharp increase not only highlights the profound impact of motor vehicle proliferation on public safety but also underscores the critical need for enhanced road safety measures and regulations that began to emerge during the early twentieth century.

Over the subsequent decades, significant strides in safety measures—such as the implementation of seat belts, airbags, comprehensive drivers' education, and heightened public awareness—contributed to a dramatic reduction in traffic deaths. These efforts have successfully decreased traffic fatalities per one hundred million vehicle miles traveled by almost 90 percent since the 1920s, reflecting a concerted push toward safer driving environments and more informed driving practices.[4] This evolution in road safety highlights a responsive shift in policy and public behavior toward the ever-present challenges posed by modern transportation technologies.

However, this progress halted around fifteen years ago in the United States, coinciding with the introduction of smartphones in 2007. Leonhardt's article points out that Americans are more

3   "Motor Vehicle Traffic Fatalities and Fatality Rates, 1899–2020," Fatality Rates and Lives Saved, Traffic Safety Annual Report Tables, National Highway Traffic Safety Administration, August 2024, https://cdan.dot.gov/tsftables/Fatalities%20and%20Fatality%20Rates.pdf.

4   Centers for Disease Control and Prevention, "Achievements in Public Health, 1900–1999 Motor-Vehicle Safety: A 20th Century Public Health Achievement," *Morbidity and Mortality Weekly Report* 48, no. 18 (May 14, 1999): 369–374, https://www.cdc.gov/mmwr/preview/mmwrhtml/mm4818a1.htm.

prone to using their cell phones while driving compared to people in other countries. This dangerous trend is exacerbated by the fact that most cars in the US are automatic, which frees up drivers' hands to use their phones. In contrast, in Europe, where 75 percent of cars are manual and using cell phones while driving is culturally frowned upon, this issue is less pronounced.[5]

This situation illustrates the paradox of progress and how new technologies, while offering numerous benefits, can also have unforeseen negative impacts. The ease of use provided by automatic vehicles and the distracting nature of smartphones have combined to create a hazardous driving environment, reversing the trend of decreasing traffic deaths. The article serves as a reminder of the need for continuous evaluation and adaptation of safety measures and public behavior in response to evolving technologies.

But traffic collisions caused by distracted drivers only happen in the news, I once thought. *They won't happen to me.*

In 2018, I was out training for an Ironman Triathlon on my bike when a distracted driver using a cell phone hit me and ran my torso over. The collision broke my back in five spots, broke my pelvis in two, fractured several ribs, punctured one lung, and broke both clavicles. A helicopter ride and several surgeries later, I was grateful to be alive, but also much more cognizant of my own cell phone use while driving. Meanwhile, the driver, an older gentleman, lost his license and went to jail. This event underscored a broader truth: Regardless of the historical or industrial context, modern technologies can have unforeseen and impactful consequences on each of us in our daily lives.

This personal experience highlights the importance of cautious navigation in the rapidly advancing field of AI similar to

---

5   Leonhardt, "Rise in Traffic Deaths."

the sweeping changes brought about by the Industrial Revolution and the widespread adoption of cell phones, the current "AI all-in" mindset presents risks of neglecting ethical considerations and failing to foresee potential unintended impacts on the broader society but also on our everyday lives. Drawing lessons from history, it becomes clear that we must adopt a proactive stance in addressing and managing the challenges posed by this new wave of technological innovation, rather than responding reactively to its consequences.

THE HIDDEN COSTS OF INNOVATION

Throughout history, technological advancements have propelled society forward, solving problems and improving lives. But progress often comes with a catch. Many breakthroughs, once celebrated for their benefits, later revealed unintended consequences—sometimes with devastating effects.

Below are ten more examples of the paradox of progress, where innovations led to unforeseen challenges:

1. **Leaded Gasoline and Public Health:** Added to gasoline to improve engine performance, tetraethyl lead caused widespread neurological damage, particularly in children.
2. **Asbestos in Construction:** Valued for its fire resistance, asbestos was later linked to deadly diseases like lung cancer and mesothelioma.
3. **Antibiotic Overuse and Resistance:** Lifesaving antibiotics have been overused, leading to the rise of antibiotic-resistant bacteria, a major global health threat.
4. **Social Media and Privacy Erosion:** While transforming communication, social media has also fueled privacy concerns, data breaches, and the spread of misinformation.

5. **DDT and Wildlife Harm:** Once a revolutionary pesticide, DDT was banned after it was found to devastate bird populations and harm ecosystems.
6. **Thalidomide and Birth Defects:** Marketed as a safe treatment for morning sickness, thalidomide tragically caused thousands of birth defects, prompting stricter drug regulations.
7. **Subprime Mortgage Crisis:** Financial innovations in subprime lending aimed at increasing home ownership instead triggered the 2008 housing market collapse.
8. **CFCs and the Ozone Layer:** Chlorofluorocarbons, used in refrigeration and aerosols, were later found to deplete the ozone layer, increasing harmful UV radiation.
9. **Automobiles and Urban Sprawl:** Cars brought convenience but also led to traffic congestion, increased emissions, and unsustainable urban expansion.
10. **Nuclear Energy and Radioactive Waste:** Although offering a cleaner energy alternative, the long-term storage challenges and potential hazards of nuclear waste have posed significant environmental and safety concerns.

Each of these examples underscores a crucial lesson: Innovation is a double-edged sword. The challenge isn't just advancing technology—it's anticipating and mitigating its unintended consequences before they spiral out of control.

A SHIFT TOWARD PROACTIVE APPROACHES

Our reactions to unintended consequences of progress are not binary but exist on a spectrum. This spectrum ranges from immediate, reactionary measures to proactive, anticipatory strategies. Historically, our responses have often been reactive, addressing issues only after significant harm has occurred.

However, with AI, we have the opportunity to shift toward more proactive approaches. Unlike past eras, we now possess the capability to anticipate and mitigate potential negative outcomes through advanced scenario planning and predictive modeling. Just as surgeons meticulously plan and discuss potential outcomes before each surgery, we can leverage AI to run simulations and explore various scenarios, allowing us to foresee and address possible consequences before they manifest. This proactive approach not only embodies the evolutionary spirit of past reforms but also emphasizes the importance of limiting unintended consequences through informed foresight and ethical deliberation.

By carefully considering potential impacts and implementing ethical guidelines and regulations early on, we can minimize negative consequences and foster a more sustainable and equitable evolution. This shift in approach requires a collective effort to prioritize foresight and ethical considerations, ensuring that the benefits of AI are realized while mitigating its risks.

As we explore the paradoxes in Part Two, I encourage you to engage critically with the historical lessons and broad impacts of technology, advocating for a thoughtful approach to AI that addresses emerging ethical challenges and prioritizes human dignity and societal well-being.

After all, AI is but a reflection of us, which we'll explore in the next chapter.

CHAPTER 2

AI IS A REFLECTION OF US

"Give me eight hours to chop down a tree and I will spend the first six sharpening the axe."

—ABRAHAM LINCOLN, 1863

Visit any website and you will find an "About Us" section. Why is that?

Because the "About Us" page is more than a webpage—it's the heart of a brand, offering a glimpse into its journey and core values. Whether it's called "Our Story," "Our Mission," or simply "About," this page draws us in, sharing the brand's vision and ambitions. That's what makes it so compelling: It offers a personal look at a company's character.

In a way, artificial intelligence is humanity's own "About Us" page. Every piece of data used to build AI comes from *us*—our knowledge, choices, innovations, and even our imperfections. AI reflects our collective experiences and values, processing information based on what we've written, done, and documented over time. It's more than a technical marvel; it's a digital mirror of our shared human story, intertwining with our progress, knowledge, and ethical struggles.

However, just as an "About Us" page tends to highlight only the roses—glossing over flaws and mistakes—AI often reflects our best intentions while obscuring the thorns. If we want to truly understand AI's impact, we must be intentional in examining not just the beauty of what it achieves but also the challenges and risks it presents. That's what this book seeks to do: to explore both the roses and the thorns, recognizing AI's potential while calling attention to the unintended consequences that may grow alongside it.

Viewing AI as an "About Us" section reshapes how we think about AI's role in our lives. It's not just about technology advancing at breakneck speed; it's about recognizing AI as a reflection of *us*—our choices, our history, and our potential. We're not merely spectators to AI's growth; we're its stewards, responsible for shaping it with care and vision, ensuring that it reflects the best of us and is fair and accessible to all.

Therefore, AI's path is inherently tied to our past, present, and future, creating a digital mosaic of human history and society.

Today, in 2025, we stand at a pivotal crossroads, different from the dawn of the Industrial Revolution because of the astounding pace of technological growth. We have the power to connect globally in ways previously unimaginable, sparking conversations on AI's future and how we, together, can guide it responsibly.

AI's "decision-making" is inherently "About Us"—our accumulated knowledge, our wins, and our lessons learned.

As a global community, we must collectively "sharpen our axes before we chop," plan ahead, and contribute to shaping the path forward for AI, focusing on ensuring its safety, fairness, and universal accessibility. In other words, we must proactively develop robust guardrails to ensure the safe and ethical use of AI, protecting against misuse while promoting innovation and equity.

In this chapter, we'll explore how global bodies like the United Nations are shaping the conversation on AI ethics, look at the complex ethical challenges in AI, and emphasize the crucial role we play in guiding AI's development. The aim is clear: to keep AI a tool that enhances humanity, expanding our capabilities without compromising who we are.

Because at its core, AI is about amplifying "us"—enhancing our human potential, not replacing it.

SETTING UP GUARDRAILS

When I say guardrails in the context of AI and emerging technologies, I mean ethical guidelines, policies, and regulatory measures designed to keep technological progress on a safe and responsible path. Just as physical guardrails on a mountain road prevent vehicles from veering off dangerous cliffs, these metaphorical guardrails aim to guide the development and application of AI in ways that respect human rights, promote fairness, and protect against harm.

We need guardrails to ensure that AI serves humanity's best interests, aligning with shared values and minimizing risks like bias, misuse, or violations of privacy.

I'd argue there is no better forum for this global discussion on guardrails than the United Nations (UN). The organization's legacy in fostering global cooperation and dialogue serves as a strong foundation for navigating the ethical landscape of AI and other future technological advancements.

Here's what I mean.

Since its inception, the UN has steadily grown into a critical platform for tackling global challenges, creating space for complex discussions and fostering cooperation across nations. This journey began with the Universal Declaration of Human Rights in 1948, a groundbreaking document that set a global standard for human rights and dignity.[6] Then, in 1968, the UN adopted the Nuclear Non-Proliferation Treaty, which took effect in 1970, marking a major step toward preventing the spread of nuclear weapons and encouraging peaceful uses of nuclear energy.[7]

By 1975, the UN had shifted its focus to gender equality and women's rights, launching the First World Conference on Women, which began a series of conferences that would help shape the global conversation on women's issues.[8] This commitment to addressing global concerns expanded further in 1982, with the United Nations Convention on the Law of the Sea (UNCLOS), which established a legal framework to guide the responsible use of the world's oceans and marine resources.[9]

As the digital age arrived, the UN adapted once more, hosting the World Summit on the Information Society in 2003 and 2005 to explore how information and communication technologies impact society.[10] This summit set the stage for the Millennium Declaration in 2000, a set of targets aimed

---

6   G.A. Res. 217 (III) A, Universal Declaration of Human Rights (Dec. 10, 1948), https://undocs.org/en/A/RES/217(III).

7   G.A. Res. 2373 (XXII), Treaty on the Non-Proliferation of Nuclear Weapons (June 12, 1968), https://docs.un.org/en/a/res/2373(XXII).

8   E. Conf. 66/34, Report of the World Conference of the International Women's Year, Mexico City, 19 June–2 July 1975 (Jan. 1, 1976), https://docs.un.org/en/e/conf.66/34.

9   A. Conf. 62/122, United Nations Convention on the Law of the Sea (Oct. 7, 1982), https://docs.un.org/en/a/conf.62/122.

10   "About: What Is the World Summit on the Information Society (WSIS)?," International Telecommunication Union, accessed March 20, 2025, https://www.itu.int/net4/wsis/forum/2025/Home/About.

at tackling critical development issues by 2015.[11] Building on that success, the Sustainable Development Goals launched in 2015, expanding to cover a wide range of social, economic, and environmental goals set to be achieved by 2030.[12]

One of the UN's biggest recent achievements came in 2015 with the Paris Agreement—a landmark step in the fight against climate change that solidified the UN's role as a leader in addressing environmental challenges on a global scale.[13]

Each of these impactful initiatives shows how the UN has evolved to stay relevant and responsive to the world's most pressing issues. Today, as we find ourselves on the brink of a new technological era with AI, the UN's role in facilitating global conversations becomes even more pertinent.

And it's a good thing they're up to the task.

GLOBAL FORUM ON AI ETHICS AND RESPONSIBILITY

The UN, through its commitment to fostering global dialogue on critical issues, took a significant step in addressing the ethics of artificial intelligence by drafting the *Recommendation on the Ethics of Artificial Intelligence* in November 2021.[14]

Written at the forty-first session of UNESCO—the United Nations Educational, Scientific and Cultural Organization—in Paris, this document marked a milestone in tackling the eth-

---

11 A. Res. 55/2, United Nations Millennium Declaration (Sept. 18, 2000), https://docs.un.org/en/a/res/55/2.

12 A. Res. 70/1, Transforming Our World: The 2030 Agenda for Sustainable Development (Sept. 25, 2015), https://docs.un.org/en/a/res/70/1.

13 FCCC. CP. 2015/L.9, Adoption of the Paris Agreement (Dec. 12, 2015), https://docs.un.org/en/fccc/cp/2015/l.9.

14 UNESCO, *Recommendation on the Ethics of Artificial Intelligence: Adopted on 23 November 2021* (UNESCO, 2022), https://unesdoc.unesco.org/ark:/48223/pf0000381137.

ical implications of AI in our rapidly changing world.[15] The forum highlighted AI's vast influence on society, the environment, and daily life, presenting it as more than a technological tool but as a transformative force reshaping human thought, decision-making, and interactions across sectors. In response, UNESCO promoted a holistic approach to AI ethics, rooted in an evolving framework that keeps pace with technological advancements and prioritizes human dignity, well-being, and harm prevention.

The conference dove deep into the entire AI lifecycle—from the initial stages of research and design to deployment, maintenance, and even the point where systems are eventually retired. This big-picture view emphasized that everyone involved—researchers, engineers, end users, governments—has a critical role in shaping AI responsibly.

The UN gets it: AI is a reflection of *us.*

One central theme from the conference was the call for global collaboration, a collective effort to make sure AI benefits all of humanity. By sharing knowledge, ethical guidelines, and best practices, the international community can build a future where AI is inclusive, transparent, fair, and accountable.

The UNESCO conference wasn't just about reacting to AI's potential risks; it was about shaping a responsible path forward. The proposed framework is a proactive step to ensure AI develops in alignment with humanity's best interests. With the principles and discussions from this event, policymakers, developers, and users now have a guide—a roadmap to help navigate the ethical landscape of AI, making sure it stays true to human values and societal needs.

This conference laid essential groundwork for a future

---

15   UNESCO, *Ethics of Artificial Intelligence.*

where AI doesn't just advance technology but also enhances the human experience.

UNDERSTANDING AI ETHICS: THE BASICS

AI ethics is a new and essential branch of ethics that tackles the moral and societal implications of artificial intelligence. It covers a range of issues, from fairness and transparency to privacy, accountability, and the broader social impact of AI.

Here's why AI ethics matters:

First, it's about preventing harm. Without ethical guidance, AI systems can reinforce biases, compromise privacy, or make flawed decisions that affect people's lives. Setting ethical standards helps minimize these risks, ensuring that AI serves humanity without causing unintended harm.

Trust is another key element. For AI to become a seamless part of society, it needs to be trusted. This can be especially challenging because certain groups may be more skeptical than others, leading to uneven adoption and benefits across different ethnic, racial, gender, and cultural backgrounds. By fostering trust through ethical practices, AI can become more widely accepted and effectively used across society. More on this in Chapter 3.

Transparency, accountability, and inclusivity are also at the heart of ethical AI. Clear insights into how algorithms work, responsibility for AI-driven decisions, and diverse, representative datasets are all essential. These practices not only build public confidence but also ensure that AI benefits everyone. To achieve this, we need ongoing engagement with communities, active bias-checking, and strong regulations to protect against misuse.

AI ethics also safeguards individual rights and liberties,

emphasizing respect for human dignity and privacy. It ensures that technology doesn't infringe on core human values.

Finally, AI ethics promotes responsible and sustainable innovation. It balances technological progress with societal well-being, aligning AI development with humanity's broader interests and environmental responsibilities.

In short, AI ethics is about more than setting rules for technology. It's about building a future where AI is a force for good, enhancing lives while honoring our shared values and rights.

HOW HUMAN VALUES SHAPE AI DEVELOPMENT

Human values are essential to shaping AI, guiding the design and application of AI systems to reflect the diverse values and ethical principles of the societies they serve. This includes a commitment to human rights, justice, equality, and democratic ideals.

A crucial concept here is Value-Sensitive Design—an approach that brings together ethicists, social scientists, experts, and everyday users to embed these values into AI systems from the start.[16]

Addressing bias and promoting fairness are critical in ensuring that AI reflects our values. AI systems must not perpetuate existing biases or create new ones. This means carefully examining training data and considering the contexts in which AI operates. Moreover, AI development should be human-centered, focusing on enhancing human capabilities and well-being, and designed to support human intelligence, creativity, and decision-making rather than replacing them.

Involving a broad spectrum of stakeholders—including

16   Steven Umbrello and Ibo van de Poel, "Mapping Value Sensitive Design onto AI for Social Good Principles." *AI and Ethics* 1, no. 3 (2021): 283–296, https://doi.org/10.1007/s43681-021-00038-3.

marginalized and underrepresented groups—is also crucial for creating inclusive AI that meets diverse needs. Additionally, adhering to international ethical guidelines and standards from organizations like UNESCO, the US, and the European Union helps maintain consistency across regions, ensuring that AI reflects shared human values.

The integration and implementation of AI ethics are vital for harnessing the potential of AI in a manner that is beneficial, fair, and aligned with humanity's core values. This endeavor calls for a concerted effort from technologists, ethicists, policymakers, everyday end users, and society at large, ensuring that AI not only serves the greater good but also respects human dignity and rights.

This holistic approach is key to realizing the full potential of AI as a force for positive change in the world.

SO, WHAT ABOUT "US"?

"Here is the catch: It's impossible to know all the ways a technology will be misused until it is used."

—WILL DOUGLAS HEAVEN, *MIT TECHNOLOGY REVIEW*, DECEMBER 2023

In the ever-evolving world of technology, AI stands as a powerful tool full of potential but also packed with complex ethical challenges. Much like a website's "About Us" page, AI reflects who we are—our values, ambitions, and achievements, but also our biases and flaws. As the *MIT Technology Review* noted in December 2023, we often only see the full impact of a technology once it's woven into *our* daily lives.[17]

---

17   Will Douglas Heaven, "These Six Questions Will Dictate the Future of Generative AI," *MIT Technology Review*, December 19, 2023, https://www.technologyreview.com/2023/12/19/1084505/generative-ai-artificial-intelligence-bias-jobs-copyright-misinformation/.

This unpredictability underscores the need for a collective effort to establish ethical guardrails and frameworks for the proper use of AI. It's not just about international discussions at platforms like the United Nations; it's about embedding this conversation into our everyday lives—in *our* coffee shops, in *our* homes, at *our* workplaces, and in *our* educational settings. From high schools to graduate schools, and in vocational and technical training, this dialogue needs to reach every corner of society to prepare us for what's to come.

AI is fundamentally about "us," which means it demands a high level of personal responsibility to understand the good, the bad, and the ugly of its rapid integration into our lives. None of "us" can afford to be asleep at the wheel this time around. The conversation about AI's future should be led by "us"—not just governments, corporations, or the UN, entities that may be out of touch with how this evolving technology impacts *our* daily lives. As such, each of us—whether we're developers, policy-makers, or everyday users—plays a part in shaping the future of AI. The choices we make now will determine if AI becomes a positive force that enhances human potential or a source of unexpected problems and inequalities.

As we stand at the edge of this new era, let's take ownership of AI's story, making sure it reflects our best ideals and aspirations. It's up to us to ensure that AI not only mirrors "us" but does so in a way that serves the greater good.

CHAPTER 3

CAN MACHINES MAKE ERRORS?

"The limiting factor is going to be the data, and making sure that the data is trusted. That the models are trusted, and that you ultimately have a trusted AI strategy."

—Juan Perez, CIO, Salesforce

I've always made a habit of showing up to class early; it gives me a chance to meet the other early birds. You'd be surprised how often I find diamonds among the crowd that turn into friendships that last a lifetime.

On one misty morning in May of 2024, I arrived at the MIT Media Lab in Cambridge, Massachusetts, a bit damp from the drizzling rain and a full hour and ten minutes early for my EMTech Digital class on emerging AI technologies. I wasn't the first one there—a middle-aged woman sat alone at the far end of the room. After grabbing a cup of hot coffee to shake off the chill, I introduced myself.

What caught my eye first was Diane Williams's three-and-a-half-foot-long dreadlock ponytail, thick as a winter scarf, snaking down the length of her back—a style that seemed both practical and poetic on this cool, wet, spring morning in Boston. Diane was a short, plump, proud Black woman in her middle years, who moved with a limp that spoke of both pain and a lifetime of resilience. She relied on a cane for support, a subtle emblem of her fortitude.

I detected a slight Caribbean island accent, and when I asked which island she was from, she broke into a deep, hearty drawn-out Jamaican-type laugh: "The Island of Manhattan." Her throaty laugh drew me in immediately; this was a person I wanted to get to know. She was real. Smart but raw, accomplished but humbled by life and time. Diane put me at ease.

As I glanced down at her name tag, I noticed she represented not only MIT but also Harvard—not a bad résumé for such a quiet, unassuming soul. As it turned out, she obtained her undergraduate degree in computer science from MIT back in the early nineties and then crossed the street to complete her graduate studies at the "other" institution—the one with the crimson flag.

Our conversation then moved beyond small talk. Diane, like me, was pivoting into AI toward the end of her last career. Her interest was AI ethics. When I asked why, she replied, "So it doesn't turn out like the last time—the internet."

I waited for a moment and let that sink in. "What do you mean?" I asked.

"Because the last tech wave left people like me out," she said, her eyes flicking toward a trio of late-thirty-something millennial men. They were all white, slightly balding, and dressed almost identically in the quintessential Silicon Valley uniform: untucked button-down shirts, black-rimmed glasses, blue blazers, urban jeans, and white sneakers. They were an amalgam of

Jobs–Zuckerberg look-alikes. If you pictured these three with a red tie, khaki pants, and patent leather shoes, you would have seen a striking resemblance to the elite school fraternity boys who were destined to become CIA or NSA agents a couple of generations ago.

"The early internet was an elite club, and I wasn't in it," Diane continued quietly, her voice tinged with a touch of regret. As she spoke, I could sense the weight of her disappointment, the feeling of having been left out of a transformative era. Her words made me reflect on the many voices and talents that were overlooked or excluded, and the missed opportunities for a more inclusive digital revolution.

"How would you have done it differently?" I asked. "How should it be done this time?"

"Well, as I mentioned, I grew up on the island," Diane began, her tone warm with nostalgia. "When you walked into a coffee shop, you'd hear twenty different languages—it was just normal to me. I remember in fifth grade, my teacher told the whole class, 'Tomorrow, don't buy lunch. Instead, bring a recipe from your culture so we can all learn from each other.'" She paused, her smile deepening at the memory. "What we got was Palestinians eating matzo balls and Brazilians eating borscht. Sure, we learned about each other's cultures, but on a deeper, more powerful level, we learned to trust each other."

She leaned in, her voice carrying the urgency of someone who knows the stakes are high. "That's how we need to build this next digital revolution. If AI is going to be as powerful as everyone says, it can't be built by just a few voices. We need everyone—every culture, every gender, every race—at the table from day one. That's how you build trust. Mistakes will happen, but trust in the machines we create is what will make or break us."

ERRORS IN JUDGMENT

Sometimes, we as humans make critical errors in judgment. These lapses can result from distractions, fatigue, biases, emotional reactions, hasty decisions, or incomplete and incorrect information.

Research suggests that humans make three to six errors per hour and approximately fifty errors per day, regardless of the task or activity.[18] In high-risk industries, human error accounts for 60–80 percent of incidents, accidents, and failures. Under conditions of extreme stress or fatigue, the error rate increases to eleven to fifteen errors per hour. In critical safety fields such as surgery, human error can have life-or-death consequences. If there is one salient lesson that I have learned after twenty-nine years of operating on humans is that 99 percent is never good enough.

But can a machine make these same mistakes too? And what happens when machines make mistakes in critical safety fields such as medicine, the military, or law enforcement?

As we progress into a world where critical decisions are increasingly made by machines rather than humans, when an artificial intelligence platform makes a critical error, what does this do to erode the trust between humans and machines?

While machines, especially those powered by AI, are often seen as more objective and consistent than humans, they are not immune to errors. This brings us to one of the many paradoxes of progress: As AI systems become more integrated into decision-making processes, the potential for such errors raises significant concerns and may erode *trust*.

---

18 Melanie Thomson, "Why Do We Tolerate Human over Machine Error?," *RM Assessment Blog*, November 20, 2018, https://blog.rmresults.com/why-do-we-tolerate-human-over-machine-error.

And without trust, the system will never be adopted. It will never progress.

Researchers at the University of Wisconsin recently conducted an experiment to compare how easily we forgive AI versus human counterparts. Initially, participants reported equal trust in both AI and human sources. However, this trust quickly changed after each made a mistake: When the AI erred, participants swiftly disregarded its advice or abandoned it entirely; conversely, when a human advisor made a mistake, there was only a 5 percent drop in participants' trust. Unlike human errors, which we may understand and forgive as part of the human condition, errors made by machines can feel more insidious and harder to justify, especially when they impact our lives in significant ways.[19]

The erosion of trust between humans and machines can have far-reaching consequences. If people start to doubt the decisions made by AI, they may resist or reject its use, even in areas where it has demonstrated clear benefits. This could slow down technological progress and hinder the adoption of innovations that could improve our lives.

Here, the paradox of progress is evident: The very tool designed to advance humanity can, when flawed, impede its own progress.

In this chapter, we examine the profound implications of AI mistakes through real-world examples like the wrongful arrest of Robert Williams, highlighting the dangers of biased data and the ethical challenges that arise from AI's misuse. We'll then explore how these incidents underscore the necessity for robust guardrails—such as stringent oversight, diverse datasets, and ethical guidelines—to ensure AI technologies serve humanity equitably, fairly, and without undermining trust in their use.

THE CASE OF ROBERT WILLIAMS

Facial recognition technology driven by AI platforms was originally designed to improve security, enhance efficiency, and streamline identification processes. While the technology offers significant benefits, such as quick identification of individuals in military security checkpoints, streamlining access control in various sectors, and even aiding in criminal or terrorist investigations, it is not without its drawbacks and potential for harm.

More importantly, flaws in any AI system can build mistrust. The following example is just one of many that showcases the potential mistrust in AI systems.

On January 9, 2020, Detroit police drove to the upper-middle-class suburb of Farmington Hills and arrested Robert Williams in his driveway while his wife and young daughters looked on in shock. Williams, a Black man, was accused of stealing watches from Shinola, a luxury store in downtown Detroit. He was held overnight in jail, bewildered and frightened.[20]

During questioning, an officer showed Williams a picture of the suspect. Williams, confused and indignant, rejected the claim. "This is not me," he told the officer. "I hope y'all don't think all Black people look alike."

The officer's chilling response: "The computer says it's you."

Williams's wrongful arrest, first reported by the *New York*

---

20   Tate Ryan-Mosley, "The New Lawsuit That Shows Facial Recognition Is Officially a Civil Rights Issue," *MIT Technology Review*, April 14, 2021, https://www.technologyreview.com/2021/04/14/1022676/robert-williams-facial-recognition-lawsuit-aclu-detroit-police/.

Times in August 2020, was based on a flawed match from the Detroit Police Department's facial recognition system.[21] This incident was not an isolated case. Additional false arrests have since been made public, mostly involving minorities, some of whom have also taken legal action.

Following these incidents, Williams decided to take a stand. He not only sued the department for his wrongful arrest but is also pushing to get the controversial technology banned. Since that time, the ACLU and the University of Michigan Law School's Civil Rights Litigation Initiative filed a lawsuit on behalf of Williams. The suit alleged that the arrest violated his Fourth Amendment rights and contravened Michigan's civil rights law.

The lawsuit demanded compensation, greater transparency about the use of facial recognition, and a complete halt to the Detroit Police Department's use of the technology, whether directly or indirectly.

The complaint argued that Williams's false arrest was a direct result of the flawed facial recognition system. It stated, "This wrongful arrest and imprisonment case exemplifies the grave harm caused by the misuse of, and reliance upon, facial recognition technology."

Although Williams advocated for a ban, this reaction doesn't address the deeper reality: Every innovation carries the potential for unintended consequences. The history of technological progress shows us that these missteps are not the end of the story—they are part of the innovation cycle. We identify flaws, learn from them, improve upon the technology, and iterate until it becomes safer and more reliable.

---

21 Kashmir Hill, "Wrongfully Accused by an Algorithm," *New York Times*, August 3, 2020, https://www.nytimes.com/2020/06/24/technology/facial-recognition-arrest.html.

Facial recognition technology is no exception.

Williams's story underscores the significant risks and potential for harm inherent in the use of facial recognition technology. The publicity of this case and others has eroded the public's trust in AI-driven facial recognition platforms even though these very technologies stand to benefit law enforcement and keep the general public safe. William's legal battle aims not only to seek justice for himself but also to prevent others from suffering similar injustices, advocating for a future where technology is used responsibly and equitably.

The wrongful arrest of Robert Williams highlights the ethical dilemma of data bias in AI, particularly in facial recognition technologies. Studies, including one by the National Institute of Standards and Technology (NIST), have shown that these technologies tend to have higher error rates for people of color and women.[22] This bias stems largely from the datasets on which the technologies are trained, which historically underrepresented minority populations and thus fail to accurately identify features across all demographic groups. Evidence from Asia underscores the importance of balanced training datasets; when facial recognition systems were trained on large numbers of Asian data points, the machines performed significantly better, demonstrating that diverse and inclusive training data can drastically reduce error rates.

Williams's case brought significant attention to the social ramifications of data bias in AI. It eroded public trust in law enforcement and its use of technology, raising concerns about privacy, civil liberties, and the potential for widespread misuse.

---

22 "NIST Study Evaluates Effects of Race, Age, Sex on Face Recognition Software," National Institute of Standards and Technology, December 19, 2019, https://www.nist.gov/news-events/news/2019/12/nist-study-evaluates-effects-race-age-sex-face-recognition-software.

Additionally, the incident highlighted systemic racial bias in AI technologies and their potential to exacerbate existing racial inequalities and discrimination, particularly within the criminal justice system. In response to these issues, there have been calls for stricter regulations and oversight of facial recognition technology, with some cities and states in the US moving to ban or limit its use by law enforcement agencies.

Following the public outcry, the Detroit Police Department acknowledged the flaws in relying solely on facial recognition technology for arrests. The case of Robert Williams prompted a broader discussion on the need for more diverse and representative datasets to reduce bias. It also emphasized the importance of implementing checks and balances, including human oversight of AI decisions, especially in high-stakes scenarios like law enforcement. Further, the case underscored the necessity of developing clear ethical guidelines and ensuring transparency in the use of AI technologies by public institutions.

Ultimately, the case was about trust.

DATA: THE FUEL THAT POWERS AI'S RACE TO EXCELLENCE

Imagine you're at the edge of a Formula One racetrack, the air pulsating with the roar of engines. This world of high-speed racing, where precision engineering meets the raw power of human ambition, serves as a perfect analogy for the intricate and dynamic field of artificial intelligence, particularly in the context of data bias and the principle that "garbage in equals garbage out."

Let's build this analogy out further to highlight how data fuels the outcome of AI.

Data as Race Fuel: Just as the performance of an F1 car is critically dependent on the quality of fuel it consumes, so too is the efficacy of AI systems reliant on the quality of data they process.

In the realm of AI, data acts as the fuel, driving the engines of machine learning and artificial intelligence forward. The purity and suitability of this data fuel can dramatically influence the AI system's output, akin to how high-octane fuel can boost a race car's speed and efficiency on the track.

If the data is biased, incomplete, or flawed, however, it's like filling a state-of-the-art AI machine with substandard fuel, inevitably leading to poor performance, unexpected results, and possibly even catastrophic outcomes during the race.

Machine Learning as the Engine: The heart of an F1 car is its engine—it transforms fuel into velocity and power. Similarly, machine-learning algorithms are the engines within AI that convert data into actionable insights, decisions, and predictions. The sophistication and design of these algorithms can greatly affect how effectively data is processed and utilized, much like how the engineering of an F1 engine determines its capability to utilize fuel for maximum performance.

Artificial Intelligence as the F1 Race Car: The race car itself represents the culmination of artificial intelligence. It's not merely about the engine (machine learning) or the fuel (data) but the integration of all components, including aerodynamics, driver input, and strategic planning, to navigate the complexities of the race. In AI, this integration involves algorithms, data processing, and decision-making frameworks working in concert to tackle complex tasks and solve problems.

Just as the success of a race car is contingent on the quality of its fuel, the integrity of the entire AI system is fundamentally tied to the quality of its *data—garbage in, garbage out.* Inaccurate, biased, or poor-quality data can skew outcomes, perpetuate inequalities, and undermine the system's reliability and fairness.

In the case of Mr. Williams, the data used to train AI facial recognition platforms were flawed and did not recognize subtleties in the faces of minorities—which led to the false arrest of an innocent man. The broader societal ramifications are that, because of this highly public case, potentially useful technology is no longer trusted by the general public.

Hence, the absolute necessity for pure and realistic data in AI cannot be overstated. The process of curating this data, ensuring its accuracy, representativeness, and fairness, is akin to selecting the finest fuel for an F1 car. It's a task that requires diligence, precision, and subject matter experts who possess a deep understanding of the intricate dynamics at play.

As we look deeper into the implications of data bias, it's critical to remember that the data fueling our AI systems is a reflection of our world, with all its complexities and biases. Ensuring the ethical sourcing and processing of this data is not merely a technical challenge but a moral imperative. Just as the performance of an F1 car on the track is a direct result of the quality of its fuel, the outcomes of our AI systems are a reflection of the data we feed them. In the pursuit of artificial intelligence that is not only powerful but also equitable and just, and most importantly trustworthy, the purity of our data is paramount.

So, where do these sources of data bias even come from?

SOURCES OF DATA BIAS

Although it is beyond the scope of this chapter to describe all forms of data bias, we will use the example of Robert Williams to illustrate how poor data can bias facial recognition software.

The sources of bias in AI are multifaceted, arising from the very data that feeds these algorithms. Prominent contributors to algorithmic bias include tarnished or incomplete data, skewed data, and a lack of up-to-date data, each playing a significant role in shaping the behavior and decisions of AI systems.

Having touched on the NIST study earlier, let's now dive deeper into its findings to explore the crucial distinction between one-to-one and one-to-many scenarios. We'll also look at another example where poor data caused poor AI outcomes—this one from Amazon. NIST Study on Facial Recognition

In 2019, the National Institute of Standards and Technology (NIST) conducted a study revealing that the accuracy of facial recognition software varies significantly across different sexes, ages, and racial backgrounds.[23] This variation largely depends on the specific algorithm used, its application, and the data it was trained on. Most algorithms showed demographic differences, meaning their effectiveness in matching two images of the same person varied among different racial-, ethnic-, or gender-based groups.

Tarnished or incomplete data refers to information corrupted by inaccuracies, errors, or biases, often reflecting historical prejudices and societal inequalities. AI systems trained on such data can inadvertently learn and replicate these biases.

The study evaluated two primary tasks of facial recognition: "one-to-one" and "one-to-many" matching. One-to-one matching is used for verification, like unlocking a smartphone or passport checking or confirming if two photos represent the same person. One-to-many matching is used for identification, determining if a person's photo matches within a larger

---

23   National Institute of Standards and Technology, "Effects of Race, Age, Sex."

database, crucial for finding persons of interest in criminal investigations or terrorist organizations.

Errors in these tasks can lead to false positives, where two different individuals are incorrectly identified as the same person, and false negatives, where the system fails to recognize two photos as being of the same person. The consequences of these errors vary dramatically. In one-to-one matching, a false negative might be a minor inconvenience, like needing a second attempt to access a device. However, in one-to-many matching, a false positive could lead to serious consequences, such as wrongful placement on a list for further investigation, potentially leading to false accusations or wrongful arrests.

The report highlighted some frightening demographic disparities in algorithm performance. For instance, one-to-one matching showed higher rates of false positives for Asian and African American faces compared to Caucasian faces, with some algorithms making mistakes ten to one hundred times more often for these groups. Interestingly, algorithms developed in Asian countries, as opposed to Western countries, did not show such pronounced disparities between Asian and Caucasian faces, suggesting that more diverse training data might lead to fairer outcomes.[24]

For one-to-many matching, the study found higher rates of false positives, particularly for African American females, emphasizing the risk of false criminal or terrorist accusations. However, not all algorithms exhibited high false positives across demographics. The most equitably trained algorithms—those algorithms trained on many faces of people from many races—also ranked among the most accurate, highlighting

---

24 National Institute of Standards and Technology, "Effects of Race, Age, Sex."

that algorithm performance varies significantly on training datasets.[25]

To mitigate the disparities in accuracy across different demographic groups, the report emphasized the need for more diverse and representative training datasets. By exposing algorithms to a broader spectrum of human features—spanning various races, ages, and genders—these datasets can help reduce bias and promote more equitable outcomes.[26]

Additionally, we need greater transparency and ethical guidelines in the development and deployment of face recognition technologies. This includes clear communication about the limitations and appropriate use of these algorithms to policymakers, developers, and end users.

Implementing rigorous testing and evaluation mechanisms, like the NIST's Face Recognition Vendor Test (FRVT) program, is crucial for continuously monitoring and improving the performance of these algorithms. FRVT is a comprehensive set of evaluations conducted by the NIST to assess the performance of face recognition algorithms submitted by vendors worldwide.

These evaluations aim to provide government agencies, industry, and the research community with unbiased, independent information about the performance of face recognition algorithms. By offering detailed insights into the accuracy, speed, and robustness of various face recognition systems, the FRVT empowers organizations to make informed decisions when selecting a face recognition system for their specific needs. Moreover, incorporating human oversight and audits in the deployment of face recognition systems, especially in critical

---

25   National Institute of Standards and Technology, "Effects of Race, Age, Sex."

26   National Institute of Standards and Technology, "Effects of Race, Age, Sex."

applications such as law enforcement, can help mitigate the consequences of algorithmic errors.[27]

AMAZON'S AI RECRUITMENT TOOL

In 2017, Amazon faced a similar issue when it was revealed that their AI recruitment tool showed bias against women.[28] The system had been trained on résumés submitted to the company over a ten-year period, predominantly from men, reflecting the tech industry's gender imbalance. This outdated dataset failed to capture recent cultural shifts, particularly the increase in women taking on roles in technology and senior management over the past decade. Consequently, the AI learned to downgrade résumés that included words like "women's," as in "women's chess club captain," or those from candidates who graduated from women's colleges.

The bias stemmed from *skewed and outdated data*, which unintentionally favored men for certain roles and perpetuated existing inequalities by assuming past hiring patterns were indicators of future success.

Upon recognizing the bias, Amazon discontinued the use of the AI recruitment tool and initiated a comprehensive review of its AI strategy for hiring. The company began exploring bias mitigation techniques, such as developing algorithms that actively identify and counteract bias in training data. They

---

27 "Guardrail Failure: Companies Are Losing Revenue and Customers Due to AI Bias," LeackStat, accessed March 20, 2025, https://leackstat.com/news-articles/guardrail-failure-companies-are-losing-revenue-and-customers-due-to-ai-bias; Reid Blackman, "If Your Company Uses AI, It Needs an Institutional Review Board," *Harvard Business Review*, April 1, 2021, https://hbr.org/2021/04/if-your-company-uses-ai-it-needs-an-institutional-review-board.

28 Maude Lavanchy, "Amazon's Sexist Hiring Algorithm Could Still Be Better Than a Human," *The Conversation*, November 1, 2018, https://theconversation.com/amazons-sexist-hiring-algorithm-could-still-be-better-than-a-human-105270.

committed to using more up-to-date, diverse, and inclusive datasets for future AI systems, ensuring broader representation of gender, ethnicity, and background.

Additionally, Amazon implemented a stronger human oversight mechanism in its recruitment process to critically evaluate AI recommendations. Amazon recognized the continued role of human subject matter experts. By taking these steps, Amazon aimed to create a more equitable and effective hiring process, highlighting the importance of addressing and mitigating biases in AI systems.

THE DUALITY OF AI TECHNOLOGY

The results of the NIST study, coupled with the real-life examples of Robert Williams and Amazon's recruitment tool, starkly illuminate the double-edged sword of AI technology. Facial recognition and AI systems promise enhanced security, efficiency in work processes, and innovative applications across various sectors, such as talent identification, job recruitment, law enforcement, and criminology. These advantages represent significant leaps forward in our ability to protect, serve, and understand communities at scale. Just as roses symbolize beauty and progress, these benefits cannot be understated.

However, the downsides, as highlighted by Williams's wrongful arrest and Amazon's biased recruitment tool, reveal a troubling undercurrent of data bias and ethical quandaries, embodying the thorns that come with this technological rose. The technology's propensity to misidentify individuals from minority groups not only raises concerns about privacy and civil liberties but also underscores the deep-seated issue of racial and gender discrimination embedded within AI systems.

Importantly, these issues erode public trust in intelligent

systems. This is especially true in "safety critical" systems like healthcare, traffic control, and law enforcement where one line of code can literally mean the difference between life or death, safety or injury, and freedom or incarceration. These incidents bring to light the urgent need to address and rectify the biases present in the datasets that train such technologies, ensuring they are as inclusive, accurate, and as representative as possible.

Williams's wrongful arrest underscores the necessity for stringent oversight, ethical guidelines, and transparent operations, especially when such technologies are employed by law enforcement and other institutions wielding significant power over individuals' lives. Similarly, Amazon's recruitment tool demonstrates how AI can perpetuate gender biases if not properly monitored and corrected. These cases highlight the need for a balanced approach that recognizes the value of technology while safeguarding against its potential harms.

The duality of AI platforms—their capacity to both innovate and infringe—demands a thoughtful examination of how we proceed. The cases of Robert Williams and Amazon's recruitment tool are not just cautionary tales but clarion calls for immediate action. They challenge us to forge solutions that prioritize equitable representation in AI training datasets, ethical foresight, and robust regulatory frameworks that ensure technologies like facial recognition and talent identification serve the greater good without compromising individual rights and societal values.

As we stand at this crossroads, the path forward is clear: Trust is imperative for valuable technology to be adopted by the general public. In order to build trust, we must integrate inclusive considerations into the DNA of technological development. No underrepresented group should be left out of the development phase. This demands a proactive stance, continuously scrutinizing and addressing the ramifications of AI during the build phase rather than retrospectively. By doing so, we can harness the incredible potential of AI while also protecting and upholding the integrity and dignity of all individuals, especially those historically marginalized.

The paradox of progress reminds us that every step forward must be accompanied by reflection and responsibility. As we embrace the transformative power of AI, we must remain vigilant in recognizing and mitigating its thorns, ensuring that our pursuit of a better future is guided by wisdom, inclusivity, and ethical consideration.

The journey with AI is a collective one. It is up to all of "Us"—scientists, engineers, academics, policymakers, and citizens—to ensure that our reliance on AI is founded on principles of integrity, equity, and respect for the full spectrum of human experience. By doing so, we can create a future where AI is trusted by all end users—by "Us."

In doing so, we develop a system that empowers everyone, not just a privileged few, shaping a world that reflects the best of what it means to be human.

Building and maintaining trust in AI requires transparency, accountability, and continuous improvement. Most importantly, as Diane alluded to in her sage-like statement about Brazilians eating borscht, in order for "Us" to trust the technology, the datasets used to train AI platforms must include characteristics, traits, and data from all users—across races, genders, age groups, countries, nations, dialects, beliefs, cultures, and financial strata. We must ensure that AI systems are designed with robust guardrails to minimize errors and biases.

Additionally, there must be clear mechanisms for addressing mistakes when they occur, ensuring accountability and transparency, and for continuously updating and refining AI systems to prevent future errors.

CHAPTER 4

TECH WARS AND GLOBAL POWER

"The nuclear arms race is like two sworn enemies standing waist deep in gasoline, one with three matches, the other with five."

— CARL SAGAN, COSMOS, 1982

On a brisk winter's day, with a storm howling outside, my eighteen-year-old son, Dane, and I sat by the crackling woodstove, feeling the warmth of the fire—but also the thrill of competition.

We were huddled over a Stratego board—a game that had captivated minds for more than fifty years with its rich blend of strategy and surprise. Like chess, Stratego demands careful planning and tactical maneuvers, but with the added twist of hidden information, as each piece's identity remains concealed until challenged. The board, a battlefield of red and blue, set the stage for two Napoleonic armies poised for combat, each led by an array of generals, spies, and soldiers, all hidden in the fog of war.

As the game unfolded, every move and countermove built a story of offense and defense. Each piece, from the cunning spy to the immovable bomb, embodied the essence of strategic warfare. Dane, with a general's sharp instincts, maneuvered his units skillfully, his eyes scanning the board, reading my tactics, planning his next assault. His youth brought boldness—a fresh, unpredictable approach that often caught me off guard and pushed the boundaries of traditional strategies.

The goal was simple for both of us: to outsmart and out-maneuver, to penetrate the opponent's lines and seize the flag while fiercely protecting our own. Every piece had a purpose; every loss mattered. It mirrored real-world conflicts, where each choice, every strategic misstep, could shift the balance of power in an instant.

Capturing the flag in Stratego wasn't just a win—it was symbolic, like stripping away an opponent's identity and sovereignty.

It reminded me of how flags have long represented the values, language, and cultural affiliations of groups, from tribes to nations. Historically, capturing an enemy's flag was as much about conquering their spirit as it was about seizing territory. Victorious armies would raise their own flags over conquered lands, often destroying the enemy's flag as a declaration of new control, an assertion of whose values and governance would rule.

The flags of the victorious were remembered as symbols of cultural strength, while those of the defeated were often forgotten, taking with them unique languages, traditions, and histories that might fade or be erased over time. This dynamic underscored the power of victory in shaping historical narratives and cultural legacy. In Stratego, each player's goal was to capture the opponent's flag—an act that mirrored the stripping away of identity and control.

As I looked at the game board, I couldn't help but draw a parallel to the present-day tech battles between global powers. In fields like AI, chip manufacturing, and green technology, the competition wasn't just about economic dominance. It was about asserting cultural and civilizational influence. The flag in Stratego, much like technological supremacy today, represented the core identity and strategic capability of each player or nation.

Lost in my thoughts on how to weave Stratego into this chapter, I made one misstep after another. As a consequence, I lost. Dane captured my flag, while I sat there realizing that trying to strategize a game and a chapter at the same time was a terrible idea.

In this chapter, we explore the complex dynamics between two nations' tech rivalry and how it mirrors a high-stakes game of Stratego, where control over critical technologies like AI, semiconductors, and green tech is more than an economic contest—it's a battle for cultural influence and global leadership. Through strategic moves and countermoves, alliances, and defensive and offensive tactics, the winner will not only shape the future of technology but also the standards, values, and geopolitical alignments that will define the twenty-first century.

THE CURRENT TECH WAR

The struggle for technological leadership has become more than a race for resources or market share; it's now a contest over which values, norms, and standards will shape the future global landscape. Nations, through their technological advancements, are attempting to capture each other's "flag" of influence, vying to set the rules of the game and establish the standards by which society operates.

The outcome of these tech battles will define which nations' cultural identities will dominate, shaping everything from international policy to individual rights and freedoms. The stakes go beyond economics, extending into the governance systems of the twenty-first century and beyond—determining whether they lean toward democracy or autocracy. This grand game of strategy underscores the immense influence of technology on the global order, echoing the intricate moves and countermoves on a Stratego board where the loss of a flag can signal a decisive defeat.

And right now, the US and China are locked in a high-stakes game of Stratego.

As both nations build their AI empires, they're not just competing for technological superiority—they're vying to establish the norms, standards, and governance that will define the digital age. While this race drives remarkable innovation, it also raises pressing concerns: Who will have access to these advancements? Will the spoils of AI exacerbate global inequalities, or will they be shared equitably? And perhaps most critically, will control over AI infrastructure pave the way for digital colonialism, where economic and political power is concentrated in the hands of a few dominant players?

In the midst of this race, flashy headlines about smartphone apps like TikTok have captured public attention, but they're mere distractions—sideline skirmishes in a far larger conflict. For instance, you probably heard about Congress's ultimatum in April 2024, giving TikTok's Chinese owners 270 days to sell their stakes or face a nationwide ban. In retaliation, China removed WhatsApp and Threads from all of its app stores, escalating tensions further.

The real battle, however, isn't about apps—it's about the chips that power them and the energy required to sustain the entire digital ecosystem.

These foundational technologies form the bedrock of this competition. Controlling chips and energy is akin to capturing the "technology flag" in this global Stratego match. Winning here isn't just about economic gains—it's about securing cultural identity, safeguarding national security, and laying claim to the future of innovation.

As we navigate this escalating tech rivalry, it's clear that the apps are just the visible tip of the iceberg. The true contest is happening beneath the surface, where control over chips and energy will determine who ultimately holds the reins in this global tech rivalry.[29]

THE TRUE BATTLEGROUND: CHIPS, GREEN TECH, AND INFLUENCE

As AI and chip production take center stage in shifting global power dynamics, the US and China are locked in an intense tech war with far-reaching consequences across sectors—from defense to the economy. Nowhere is this competition more heated than in chip manufacturing, a foundational industry supporting the world's data-processing needs for AI and other advanced technologies. Historically, the US has led in semiconductor development, but China's strategic focus on monopolizing key mineral assets and ramping up its chip production signals a pivotal shift in the tech landscape.

This competition isn't just about economic gains; it's about securing a seat at the global power table. Semiconductors, which power everything from smartphones to industrial sys-

29 Hideki Tomoshige, "Innovation Lightbulb: Innovation Competition in Chip Design Between the U.S. and China," Center for Strategic and International Studies, February 21, 2025, https://www.csis.org/analysis/innovation-lightbulb-innovation-competition-chip-design-between-us-and-china.

tems, are the backbone of the digital age. A chip shortage can bring industries to their knees, underscoring the vital role this technology plays in global supply chains and economic stability. Without robust chip production, resilience is impossible.

Adding to the stakes is green technology, the rising star of sustainable growth. As AI drives higher energy demands, clean power sources like solar, wind, and energy-efficient systems become critical. Nations that lead in green tech aren't just gaining a trade advantage—they're insulating themselves from the volatility of fossil fuel markets. This translates into greater economic stability and an influential voice in shaping the future of global energy policies.

Integrating AI with advanced semiconductor technology is key to optimizing green technologies. AI can enhance the efficiency and effectiveness of renewable energy systems through better forecasting, adaptive control systems for smart grids, and the automation of energy-efficient processes. Likewise, advancements in chip production are essential for creating the sophisticated sensors and controllers used in today's green technologies, from electric vehicles to smart energy solutions. These components aren't just crucial to their operation but also to making these technologies scalable and affordable worldwide.

As the tech wars evolve, green technology is set to become a pivotal arena. Nations leading in green tech will likely steer global energy policies and shape international environmental strategies. This competition isn't just about technological innovation; it's about securing a stake in what may become the backbone of the future global economy. The race for green dominance promises to reshape global supply chains, influence international policy alignments, and even forge new alliances based on shared interests and standards in green technology.

Leadership in this field holds the key not only to environmental stewardship but also to economic and geopolitical power in the twenty-first century.

The production of advanced semiconductors isn't just about economic growth, either—it's about controlling the flow of information. These chips power data centers, digital communication networks, and cybersecurity systems, giving nations that lead in this technology a strategic edge in geopolitics. Similarly, green technology advancements play a crucial role in energy management. Innovations like smart grids optimize energy use and make it possible to seamlessly integrate renewable power into existing infrastructures, enhancing both sustainability and reliability.

Leadership in these fields means more than technological prestige; it secures a nation's role in managing essential global resources. Control over information flow and energy production strengthens a country's position on the international stage, turning technological prowess into geopolitical power.

As this tech rivalry intensifies, the US and China are deploying a mix of strategic moves and countermoves, blending defensive measures with bold initiatives to secure dominance. This isn't just a battle for technological superiority—it's a race to define the rules and norms of the future, with chips and green tech as the critical battlegrounds.

Let's explore how these strategies are unfolding and shaping the global tech hierarchy.

America's Strategic Moves

To counter China's push for dominance in chip manufacturing, the US government has ramped up efforts to bring chip production back home. Recently, Washington announced significant subsidies to boost domestic semiconductor manufacturing. On

April 8, 2024, a $6.6 billion subsidy was approved for Taiwan Semiconductor Manufacturing Company (TSMC) to establish three new fabs in Arizona.[30] Just days later, Samsung received $6.4 billion to build new fabs in Texas, while Micron, a US company, secured $6.1 billion for semiconductor plants in New York and Idaho.[31] These investments fall under the broader CHIPS and Science Act of 2022, a $52 billion initiative designed to spur US-based chip production and train skilled workers, a critical move to ensure the US remains competitive.[32]

Alongside its focus on chip production, the US is also prioritizing green technology, as highlighted by the Inflation Reduction Act (IRA) of 2022. This $369 billion green subsidy package promotes domestic production of environmentally friendly equipment through tax credits, reinforcing a strategic shift toward sustainable technology. However, even as these initiatives take shape, the US continues to maintain significant tariffs on Chinese solar panels and electric vehicles (EVs), set at 50 percent and 100 percent, respectively—highlighting the ongoing tensions and competitive stance against Chinese imports in these crucial sectors.[33]

These mixed offensive and defensive policy measures reflect a comprehensive US strategy to secure technological indepen-

---

30 David Shepardson and Stephanie Kelly, "TSMC Wins $6.6 Bln US Subsidy for Arizona Chip Production," Reuters, April 8, 2024, https://www.reuters.com/technology/tsmc-wins-66-bln-us-subsidy-arizona-chip-production-2024-04-08/.

31 Jacob Taylor, "Huge CHIPS Grants Awarded to TSMC, Samsung, and Micron," FYI (blog), American Institute of Physics, April 19, 2024, https://ww2.aip.org/fyi/huge-chips-grants-awarded-to-tsmc-samsung-and-micron.

32 Lamar Johnson, "Biden Ends Slog on Semiconductor Bill with Signature," Politico, August 9, 2022, https://www.politico.com/news/2022/08/09/biden-ends-slog-on-semiconductor-bill-with-signature-00050530.

33 David Lawder, "US Locks in Steep China Tariff Hikes, Some Industries Warn of Disruptions," Reuters, September 13, 2024, https://www.reuters.com/business/us-locks-steep-china-tariff-hikes-many-start-sept-27-2024-09-13/.

dence and economic security, emphasizing chip production and green technology as critical arenas in the escalating global tech war.

China's Countermoves

In response, China has made significant offensive strides in green sectors critical to sustainable development, notably seizing dominance in solar panel manufacturing, battery production, and EVs. BYD, a Chinese company, is now challenging Tesla for the title of the world's largest EV manufacturer, reflecting China's strategic positioning to capture the growing global demand for sustainable solutions.

China has dramatically increased its investment in science and technology, with its 2024 budget for this sector reaching 371 billion yuan (about $52 billion USD), a 10 percent increase from the previous year.[34] This follows substantial expenditures of 1.08 trillion yuan ($152.2 billion USD) in 2023 and 650.8 billion yuan ($91.7 billion USD) specifically for high-tech industry R&D in 2022. Additionally, enterprise R&D spending reached 2.39 trillion yuan ($337 billion USD) in 2022, showing an 11 percent rise from 2021.

This aggressive funding is part of China's broader strategy to transform its economy and circumvent Western technological restrictions. The nation has tightened state control over key areas, such as quantum computing and semiconductor manufacturing, by adopting a comprehensive "whole nation" approach that mobilizes the country's resources to advance technology and overcome US barriers.

---

34 Smriti Mallapaty, "China Promises More Money for Science in 2024," *Nature*, March 8, 2024, https://doi.org/10.1038/d41586-024-00695-4.

In comparison, the US invested $806 billion in tech R&D in 2021, while China's spending reached $668 billion, highlighting China's rapid progress in R&D investment over the last two decades. This intensifying dynamic underscores the fierce competition between these superpowers in the global tech race.[35]

Defensive and Offensive Strategies in Play

Defensively, the US strategy includes limiting China's influence domestically and encouraging allies to adopt similar stances. Measures under consideration from Treasury Secretary Janet Yellen include restrictions on Chinese green tech imports, such as solar panels and EVs.[36] Meanwhile, Chinese companies are finding creative ways to navigate these barriers, exploring joint ventures with US companies or setting up manufacturing in countries with favorable trade agreements.[37]

Despite these barriers, China's massive domestic market and reach in regions outside the Western alliance bolster its green tech sectors. According to a 2024 report by *Bloomberg*, China installed more solar panels in 2023 alone than currently exist in the US, signaling China's aggressive expansion and competitive edge in solar energy and beyond.[38] While the US maintains an advantage in semiconductor production, China's advancements

---

35 Chris Buckley, "What Slowdown? Xi Says China Must Win the Global Tech Race," *New York Times*, March 11, 2025, https://www.nytimes.com/2025/03/11/world/asia/china-xi-trump.html.

36 Karen Gilchrist and Ruxandra Iordache, "Yellen Says She Won't Rule Out Possible Tariffs on China's Green Exports," CNBC, last updated April 9, 2024, https://www.cnbc.com/2024/04/08/yellen-says-she-wont-rule-out-possible-tariffs-on-chinas-green-exports.html.

37 Shameen Prashantham and Lola Woetzel, "How Chinese Companies Expand Globally Despite Headwinds," *MIT Sloan Management Review*, February 3, 2025, https://sloanreview.mit.edu/article/how-chinese-companies-expand-globally-despite-headwinds/.

38 "China Added More Solar Panels in 2023 Than US Did in Its Entire History," *Bloomberg*, January 26, 2024, https://www.bloomberg.com/news/articles/2024-01-26/china-added-more-solar-panels-in-2023-than-us-did-in-its-entire-history?srnd=green&sref=Oz9Q3OZU.

and market foothold in green technologies position it as a formidable contender in this evolving tech rivalry.

In the past, power often meant military and territorial dominance. Today, the keys to global hegemony lie in information, technology, and energy—pillars that are increasingly critical in the race for control in the modern era.

THE WHY

While much of the focus in the US–China tech war revolves around chip fabs, green tech investments, and economic dominance, the deeper layer—the question of who shapes the values and norms of our global future—remains the critical issue. This isn't just a battle for technology; it's a fight over cultural and civilizational influence, and that's why it matters. Understanding this layer is essential to grasping the true stakes of these tech battles and why they deserve our attention—even if it rarely makes headlines.

Dr. Evan Ellis, a research professor at the US Army War College's Strategic Studies Institute, explores how technological dominance can shape the very fabric of global civilization—especially in terms of cultural and civilizational influence. Ellis emphasizes that controlling critical resources like information and energy—and the infrastructures that support them—significantly impacts the cultural, social, and political dynamics of nations. In today's world, where technology is essential to both daily life and national security, countries leading in these key areas hold the power to influence global norms, values, and even the overall trajectory of civilization itself.[39]

---

39   R. Evan Ellis, "Is China Cornering the Green Energy Transition in Latin America?,"
      Diálogo Américas, February 28, 2024, https://dialogo-americas.com/articles/
      is-china-cornering-the-green-energy-transition-in-latin-america/.

Ellis suggests that if either of the tech superpowers—the US or China—gains dominance in essential technologies for information processing (like AI and chip making) and sustainable energy (green technologies), they will have the power to set international standards, governance models, and potentially even cultural practices worldwide. For instance, the country that sets the rules for AI ethics or green technology standards could influence global policies in a way that spreads its own cultural values and priorities around the world.

Ultimately, the "flag" of the victorious nation—symbolizing its cultural identity and achievements—will be celebrated and remembered, while the influence of the loser risks being overshadowed or even forgotten. This highlights the intense stakes of the global power game, where technological leadership determines which civilization's values and practices will dominate the global stage.

This is the "why" behind the billions poured into the tech wars by each superpower.

Now when you read a news story about Congress pouring money into AI, you'll understand why. The decisions made today will determine whether the US retains its ability to shape global standards, ensuring that the values embedded in emerging technologies reflect democratic ideals rather than authoritarian controls. These legislative actions aren't just about funding innovation—they're about safeguarding the principles that define a free and open society.

Whoever dominates technology holds immense ethical and cultural power. It's not just about innovation—it's about shaping the rules of surveillance, privacy, human rights, and environmental responsibility. If one nation's values take over through its technological influence, it could overshadow other cultures and ways of life, creating a kind of technological imperialism. This raises tough questions: How do we protect diversity, autonomy, and the right of each culture to define its own future without being forced to follow someone else's blueprint?

PAST STRATEGIC MOVEMENTS

Looking back at history helps us understand how past victories shaped the global order, setting the stage for today's tech rivalry. Just as the United States once leveraged technological dominance to define the cultural and economic blueprint of the modern world, China is now vying to do the same, challenging the established balance of power.

Historically, wars—both physical and technological—have defined the trajectory of civilizations, with the victors setting the cultural and economic tone for generations. Following World War II, the United States emerged as the dominant global force, leveraging its ethnocentric wins to shape the world order as we know it. From 1945 onward, American innovation became synonymous with progress, as the nation pioneered technologies that would fundamentally alter the global landscape. The development of the computer, the internet, and, more recently, electric and autonomous vehicles cemented the US's position as a technological and cultural leader.

This dominance wasn't accidental—it was the result of deliberate strategic planning. The US government's strong support for research and development created a robust framework for innovation. Through initiatives like the GI Bill, which expanded education and technical expertise, and the establishment of institutions such as NASA and DARPA, the US invested in technologies that not only advanced military capabilities but also fueled civilian progress (more on this in Chapter 11). Silicon Valley, Seattle, and other tech hubs flourished during this period, blending industry, academia, and government partnerships to create a thriving innovation ecosystem.

The ripple effects of America's technological victories post-1945 are still felt today. The global adoption of the internet, for example, has largely been guided by American values of openness and free communication—ideals that reflect the United State's cultural influence. Similarly, standards in computing, telecommunications, and now green technology are built on frameworks established during this period of American ascendancy. These victories underscore the precedent: Whoever dominates in innovation controls not just the tools of the present but also the cultural blueprint for the future.

Now, under Xi Jinping's leadership, China is looking to rewrite that blueprint. Aggressively pursuing technological self-sufficiency, China is building a *comprehensive indigenous innovation ecosystem* designed to rival and surpass the United States in key sectors.[40] By investing heavily in quantum computing, AI, and biotechnology; reforming STEM education; and fostering homegrown intellectual property, China is challenging the precedent set by the US decades ago.

The stakes are enormous. These strategic moves are not just about securing technological supremacy but about defining the cultural and ethical norms that will guide the next wave of global innovation. As with past victories, the nation that emerges ahead will likely shape global supply chains, governance standards, and the technologies that influence everyday life.

---

40 Owen J. Daniels, *CSET Analyses of China's Technology Policies and Ecosystem: The PRC's Domestic Approach* (CSET, September 2023), 14–18, https://cset.georgetown.edu/wp-content/uploads/20230035_The-PRCs-Domestic-Approach.pdf.

ALLIANCES AND COUNTER-ALLIANCES

Just as World Wars I and II forced nations to align with one side or the other, the current US–China tech rivalry is pushing countries into uncomfortable choices. This dynamic—reminiscent of a bygone era of rigid alliances—underscores how technological competition can reshape the global order. Nations are now being compelled to navigate a geopolitical landscape where neutrality is increasingly untenable, driven by fierce battles over semiconductors, AI, and sustainable technologies.

This tech war is doing more than determining the next leaders in innovation—it's embedding a new era of techno-nationalism as a global norm.[41] Countries must weigh economic dependencies, security needs, and long-term strategic goals as they decide which side to align with. For many, these decisions are not made lightly; they involve recalibrating trade policies, strengthening defense partnerships, and reconsidering national priorities.[42]

Alliances are shifting, much like they did during the world wars, but this time around, the battleground is technological. For instance, the CHIPS Act and similar US initiatives are designed not only to bolster domestic industries but also to entice allied nations into exclusive technology-sharing agreements. Similarly, China's Belt and Road Initiative incorporates technology transfers and partnerships that bring developing nations into its sphere of influence.

Countries caught in the middle—many of which prefer not to take sides—face mounting pressure to pick a camp. For

41 Kate Lamb, "The Rise of Techno-Nationalism—And the Paradox at Its Core," World Economic Forum, July 3, 2019, https://www.weforum.org/stories/2019/07/the-rise-of-techno-nationalism-and-the-paradox-at-its-core/.

42 Anu Bradford, *Digital Empires: The Global Battle to Regulate Technology* (Oxford University Press, 2023), 183–220.

them, this dynamic isn't just inconvenient; it's destabilizing. Economic policies, defense strategies, and diplomatic relationships are all influenced by the tectonic shifts in global power dynamics spurred by the US–China tech rivalry.

History shows that when nations are forced to choose sides, global tensions escalate, often leading to unintended and far-reaching consequences. The fact that we are seeing such rigid alignments forming again is not a good sign. It signals a world increasingly divided by technological allegiances, one that risks heightening competition at the expense of collaboration. The parallels to the past are striking—and they serve as a sobering reminder of the stakes involved in the ongoing tech wars.

FUTURE BATTLE PLANS: WHAT THE US SHOULD BE DOING

The Atlantic Council Strategy Paper Series dives straight into the heart of the US–China tech rivalry, suggesting ways to protect the US's position as a global tech leader.[43] Published by one of the leading US think tanks on global security, these papers offer strategic insights to help policymakers, industry leaders, and the public navigate today's complex global challenges.

The following is a brief overview of their key recommendations: a battle plan, so to speak.

To stay ahead, the strategy stresses that the US must address long-standing gaps in investment and step up to meet the pace of global innovation. Among its key recommendations is a return to strong public funding for research and development,

---

43   Peter Engelke and Emily Weinstein, "Global Strategy 2023: Winning the Tech Race with China," Atlantic Council Strategy Paper Series, June 27, 2023, https://www.atlanticcouncil.org/content-series/atlantic-council-strategy-paper-series/global-strategy-2023-winning-the-tech-race-with-china/.

the lifeblood of scientific discovery and tech advancement. From AI to life-saving healthcare innovations, reinvesting in R&D keeps the US competitive in fields that impact us all.

The paper also calls for bolstering STEM education across the board, creating a skilled workforce ready to tackle tomorrow's challenges. But it doesn't stop there—the strategy urges the tech sector to reflect America's diverse demographics, cultivating a broader range of ideas and innovations. By doing so, it aims to keep American innovation inclusive and forward-thinking, rooted in fresh perspectives.

On the talent front, the strategy emphasizes retaining the world's brightest minds while sharpening strategic decision-making across government agencies, ensuring we make the smartest, most informed moves on the global stage.

To reinforce this leadership, the plan recommends specialized science and technology intelligence units to guide policy and help US companies stay competitive internationally. It also advocates for protective measures that keep our technological edge without alienating our allies.

And finally, the strategy doesn't see the tech rivalry as pure cutthroat competition—it proposes balancing this race with diplomacy. Building strong global alliances and engaging China strategically on shared issues, like climate change and public health, could mitigate conflict while reinforcing US leadership.

In short, these "battle plans" are more than just policies. They're a call to action to secure a future where US values, innovation, and cooperation lead the world in ways that impact every citizen, from the economy to our daily lives.

CLOSING THOUGHTS: OUR PATH FORWARD

The Sino–American tech wars offer a stark reminder of the dual nature of technological advancement: the power to drive transformative progress and the potential to deepen divisions. This rivalry, while spurring rapid innovation, forces us to confront the paradox of progress—how the same tools that can revolutionize our world can also threaten its stability.

On the one hand, this competition accelerates breakthroughs in AI, semiconductor manufacturing, and green technology—advances that could redefine healthcare, education, and sustainability. The possibilities for improving lives globally are immense, with innovation driving solutions to some of humanity's greatest challenges.

On the other hand, the pursuit of tech dominance comes with grave risks. A world divided by techno-nationalism could see growing inequalities and heightened tensions, where power is determined by control of technology rather than shared human values. If cooperation gives way to competition, this divide could lead to a fractured global landscape—and, in the worst-case scenario, outright conflict.

The themes from the previous chapter, "Bridging Digital Divides," echo here, highlighting the need for an ethical, inclusive approach to technology that considers its global implications. The tech wars perfectly illustrate the nonlinear nature of progress: AI's potential to uplift humanity contrasted with the dangers of monopolization and misuse.

To ensure technology serves its true purpose—the betterment of humanity—we must prioritize diplomacy and international cooperation. The conversation must be about "Us." By fostering global dialogue and creating multilateral frameworks, we can mitigate risks, bridge divides, and maximize the shared benefits of these advancements.

The future doesn't have to be a zero-sum game. With collaboration and ethical leadership, progress can belong to all of us.

These four chapters in Part One have provided the necessary context for understanding AI's role in shaping our world. We've explored how technological progress has always come with unintended consequences, how AI reflects both the best and worst of humanity, how trust is a critical factor in its adoption, and how global competition shapes the development and control of these technologies. With this foundation, we now turn to the paradoxes of AI—examining the ways in which these technologies offer both promise and peril, solutions and new dilemmas. Each chapter in Part Two will explore a distinct paradox, revealing the roses and thorns of AI's progress and the difficult questions we must grapple with as we shape the future.

A SMALL SELECTION OF AI'S DEFINING PARADOXES

CHAPTER 5

AUTONOMOUS VEHICLES: THE ROAD AHEAD

"The way I see it, if you want the rainbow, you gotta put up with the rain."

—DOLLY PARTON

On a brisk, gray winter's day in 1886, a fifty-two-year-old German man in a thick wool coat approached the Berlin patent office. Hands trembling, not just from the cold but from excitement, Carl Benz was about to present the world with a game changer. For a decade, he had poured his heart and soul into this invention—a machine he hoped would usher in a new era. As he handed over the patent papers, stamped with the number DRP-37435, he knew that what he'd created wasn't just a machine.

It was the first automobile, a device that would redefine human mobility and open up a world of possibilities.[44]

Benz's creation, the Benz-patent Motorwagen, packed a punch with its four-stroke 58.3-cubic-inch single-cylinder engine, hinting at a future full of exploration and connection. It wasn't just a hunk of metal; it was a promise, a glimpse of a post-horse-and-buggy world, like a rainbow after a storm, radiating hope and endless potential.

Two years later, it was Carl's wife, Bertha Benz, who would prove what lay at the end of that rainbow. With her two sons in tow, Bertha took the Motorwagen on an ambitious 180-kilometer journey from Mannheim to Pforzheim. This wasn't just a test of the car's endurance—it was a declaration that the future was here. As they passed through villages and countryside, the idea of mobility transformed with every mile, turning roads into highways of possibility and showing that people were no longer limited by where their feet could carry them.

When Bertha and her sons returned, news of their journey spread, and imaginations were ignited. Suddenly, the automobile wasn't just a novelty; it was a vehicle for dreams, taking people from "where we are" to "where we want to go." Roads became pathways to new experiences and aspirations.

Of course, every rainbow has its rain. As the automobile became mainstream, the risks appeared just as clearly as the rewards. The freedom it promised came with new hazards, as speeding cars replaced slower, predictable horse-drawn carriages. Accidents became a sobering reminder that progress has a price, prompting society to implement safety rules, traffic

---

44   "The First Automobile: 1885–1886," Mercedes-Benz Group, accessed March 20, 2025, https://group. mercedes-benz.com/company/tradition/company-history/1885-1886.html.

laws, and road designs to help prevent the dangers that came with this exciting, unpredictable machine.

The automobile evolved, and so did society's relationship with it. What began as a marvel became an essential part of life. This journey of the automobile, like the fleeting beauty of a rainbow, reveals a powerful truth about progress: Each new technology, no matter how incredible, arrives with both upsides and unexpected downsides. The car's story mirrors the broader tale of technological innovation—each breakthrough propels us forward but also reminds us of the balance needed between ambition and responsibility.

While autonomous vehicles promise a future of safer roads, fewer accidents, and enhanced mobility, they come with a paradox: As AI takes the wheel, we relinquish control—raising ethical, infrastructural, and legal dilemmas. Who is responsible when an AI-driven car makes a fatal mistake? How do we integrate these vehicles into societies built for human drivers? In this chapter, we'll explore how each advancement in transportation brings both progress and new challenges, redefining mobility while demanding careful, responsible implementation.

Paradox: AI-powered vehicles promise safer, more efficient roads, but they risk creating new ethical, logistical, and regulatory challenges.

The roses: Reduced accidents and traffic efficiency.

The thorns: Ethical dilemmas (e.g., who does the car save in a crash?), job loss for drivers, and regulatory hurdles.

FROM FARM ROADS TO AUTOPILOT

My first encounter with an autonomous car and AI-driven traffic efficiency systems was nothing short of a comedic juxtaposition of past and future, set against the backdrop of busy Santa Monica, California.

I found myself in the passenger seat next to Negin Bemanzadeh, a high-powered CEO of EEE Corp Group in London, handling both the driver's seat and our meeting with effortless precision. Though she wasn't touching the wheel, she was in control in every other sense—eyes locked on her laptop, fingers flying over the keyboard, while the sleek Dual Motor Tesla Y model navigated the chaotic Los Angeles traffic on its own, weaving in and out of lanes like it had a mind of its own.

Meanwhile, I was hanging on to the safety handle for dear life, teeth clenched, bracing for an accident that never came.

I was light-years from the 1970s farm roads of rural Pennsylvania where I grew up, where driving required both hands, full attention, and nerves of steel. Now, here I was, in what felt like a scene straight out of *The Jetsons*, watching a car handle itself while its "driver" barely glanced at the road.

The irony wasn't lost on me. While Negin, the epitome of modern efficiency, was making the most of her day, I was paralyzed with skepticism, barely able to focus on our meeting. It was a moment of self-deprecating realization—my rural roots hadn't exactly prepared me for this leap into the autonomous age.

But as I clutched the handle, marveling at the gap between where we started with Benz's Motorwagen and where we've come, I couldn't help but wonder: What kind of high-tech, rainbow-hued future lies just around the bend?

What problems could it solve?

OUR CURRENT AUTO PROBLEM

Every day, our roads bear witness to a silent crisis. In the United States alone, 42,795 people were killed in motor vehicle crashes in 2022—a number that carries a heavy emotional toll, but also a massive financial one.[45] The National Highway Traffic Safety Administration (NHTSA) estimated the annual economic cost of auto accidents to be $340 billion. If quality of life factors are added in, the total value of societal harm from motor vehicle crashes in 2019 alone was nearly $1.4 trillion.[46]

But the damage doesn't stop there.

Every day, Americans lose precious time stuck in traffic. The average American commuter spends about twenty-seven minutes on their one-way commute, as reported by the US Census Bureau 2023 report.[47] Approximately 10 percent of commuters reported a daily one-way commute of at least one hour. This congestion not only costs Americans billions of dollars each year in lost productivity, but it also wastes an estimated three billion gallons of carbon-emitting fuel annually, further compounding the economic and environmental toll of traffic congestion.

In Los Angeles, drivers spent more than 100 hours, or 12.5 potentially productive 8-hour workdays a year in gridlock. The Bay Area of San Francisco is worsening each year and quickly approaching Southern California standards, reporting about

45 "NHTSA Estimates for 2022 Show Roadway Fatalities Remain Flat After Two Years of Dramatic Increases," National Highway Traffic Safety Administration, April 20, 2023, https://www.nhtsa.gov/press-releases/traffic-crash-death-estimates-2022.

46 Tanya Mohn, "Traffic Crashes Cost U.S. $340 Billion a Year, That's $230 in Taxes for Every Household," *Forbes*, January 16, 2023, https://www.forbes.com/sites/tanyamohn/2023/01/16/traffic-crashes-cost-us-340-billion-a-year-thats-230-in-taxes-for-every-household/.

47 "Selected Economic Characteristics: Mean Travel Time to Work (Minutes)," American Community Survey, United States Census Bureau, accessed March 20, 2025, https://data.census.gov/table/ACSDP1Y2023.DP03.

94 hours or 11.75 entire days of wasted time in traffic per person per year.

The national average is just over fifty hours per year per person with cities like Boston; Washington, DC; and Chicago leading the charge equating to a combined economic loss to the US of $120 billion annually due to wasted time in traffic. In an age where fossil fuels are a limited, dangerous, and costly commodity, this inefficiency of time added approximately $134 per year of added "idle time" fuel costs per person.[48]

The cost of our current traffic and safety problems is staggering, not just in terms of dollars lost or time wasted, but in lives impacted and opportunities missed. Clearly, our transportation system is overdue for a change that prioritizes both safety and efficiency. This is where AI offers tantalizing potential. From streamlining traffic flows with predictive analytics to reducing accidents through autonomous vehicles, AI-driven solutions could transform our roads from bottlenecks of frustration and risk into smoother, safer pathways.

But with this promise come significant challenges. As we explore the roses and thorns of AI in transportation, we'll examine how these technologies could revolutionize our commutes—while also bringing forth ethical questions, infrastructural demands, and unforeseen consequences.

48 Charles Degliomini, "Traffic Congestion Costs Commuters Valuable Time and Has a Heavy Economic Cost, Costing $120 Billion a Year—Is This Company's A.I. the Solution?," Access Newswire, May 26, 2023, https://www.accessnewswire.com/newsroom/en/publishing-and-media/traffic-congestion-costs-commuters-valuable-time-and-has-a-heavy-economic-cost--757275.

THE ROSES
HOW CAN AUTONOMOUS CARS HELP?

The journey toward autonomous cars, or self-driving vehicles, is a remarkable tale of innovation, tracing back to early experiments in the mid-twentieth century. Once mere figments of science fiction, these vehicles have, over decades of research and development, progressively approached reality.

This trajectory was notably accelerated by events like the Defense Advanced Research Project Agency (DARPA) Grand Challenges in the early 2000s, which catalyzed advancements in autonomous navigation and robotics, laying the foundation for today's advanced automated travel.[49] The introduction of autonomous vehicles marks a significant leap forward in the evolution of transportation, envisioning a safer, more fluid, and accessible future for commuters. These vehicles, by virtue of their design, are adept at reducing traffic-related accidents significantly, chiefly those that stem from human errors, by leveraging their perpetual monitoring and swift response capabilities.

Currently, for every person killed in a motor vehicle crash in the US, eight are hospitalized, and one hundred are treated and released from the emergency rooms. McKinsey's research suggests a profound impact of autonomous vehicles (AVs) on road safety, indicating a potential reduction in traffic fatalities by up to 90 percent. Such a decrease would remarkably shift car accidents from being the second-leading cause of death in the US to the ninth. Furthermore, AVs present a significant

---

49 "The DARPA Grand Challenge: Ten Years Later," Defense Advanced Research Projects Agency, March 13, 2014, https://www.darpa.mil/news/2014/grand-challenge-ten-years-later.

economic benefit, potentially saving over $190 billion in costs associated with fatal accidents.[50]

Despite these advantages, public acceptance varies. A study from 2018 highlighted that 73 percent of Americans expressed apprehension about riding in an AV.[51] Contrastingly, 2023 research showcases the enhanced safety of self-driving cars, demonstrating a crash rate of 23 crashes per million miles, significantly lower than the 50.5 crashes per million miles rate for human drivers. Additionally, AVs reported lower injury rates at 0.06 injuries per million miles and a remarkable figure of zero fatalities per million miles, compared to human drivers' 0.24 injuries per million miles and 0.01 fatalities per million miles, respectively.[52]

In synthesizing these insights, it's clear that autonomous vehicles hold the potential to transform road safety, traffic efficiency, and economic savings substantially. Yet, bridging public skepticism and effectively integrating these technologies into societal norms remains a pivotal challenge toward realizing the full benefits of autonomous vehicles.

Further enriching the urban landscape, autonomous vehicles have the potential to minimize the sheer volume of cars on the roads. By integrating ride-sharing and car-sharing models, especially in densely populated urban areas, it becomes feasible to serve the transportation needs of the populace with significantly fewer vehicles. This approach not only reduces vehicular density and, by extension, energy consumption and environmental

---

50   Michele Bertoncello and Dominik Wee, "Ten Ways Autonomous Driving Could Redefine the Automotive World," McKinsey & Company, June 1, 2015, https://www.mckinsey.com/industries/automotive-and-assembly/our-insights/ten-ways-autonomous-driving-could-redefine-the-automotive-world#/.

51   Ellen Edmonds, "AAA: American Trust in Autonomous Vehicles Slips," AAA Newsroom, May 22, 2018, https://newsroom.aaa.com/2018/05/aaa-american-trust-autonomous-vehicles-slips/.

52   Wall-Y, "Self-Driving Cars Are Safer Than Human Drivers, Study Shows," Warp News, October 6, 2023, https://www.warpnews.org/transportation/self-driving-cars-are-safer-than-human-drivers-study-shows/.

impact but also scales down the expenses tied to road construction and maintenance while diminishing noise pollution and the overall environmental footprint of urban transportation. Additionally, the capability of autonomous vehicles to autonomously coordinate among themselves for speed adjustment and route optimization leads to smoother traffic flow and shorter commute times, thereby enhancing the efficiency of the transportation system as a whole. This collaborative mechanism among vehicles promises to alleviate traffic congestion and elevate the overall productivity of the transportation network, setting the stage for a more streamlined and effective commuting experience.[53]

TRAFFIC EFFICIENCY PLATFORMS

Imagine pulling up to an intersection and seeing a new light—a white one. But this isn't just any light; it's the signal that autonomous cars have taken charge of the flow. Instead of the usual stop-and-go routine, this white light means you can relax a bit and simply follow the car in front of you. No second-guessing who goes next or trying to beat the light. It's traffic control, but smarter, smoother, and built for efficiency. As Hajbabaie put it to the Associated Press, "When we get to the intersection, we stop if it's red, and we go if it's green. But if the white light is active, you just follow the vehicle in front of you." Think of it as the ultimate way to breeze through intersections, with the robots keeping things moving.[54]

---

53   David Schrank et al., *Urban Mobility Report 2019* (Texas A&M Transportation Institute, August 2019), 14, https://static.tti.tamu.edu/tti.tamu.edu/documents/umr/archive/mobility-report-2019.pdf.

54   Brie Stimson, "Self-Driving Cars Could Lead to a Fourth, White Traffic Signal—Or No Signals At All: Researchers," Fox Business, May 11, 2024, https://www.foxbusiness.com/technology/self-driving-cars-fourth-white-traffic-signal-no-signals-researchers.

This intriguing idea promises a future where human drivers can seamlessly integrate into a coordinated, self-regulating flow of autonomous vehicles. By simply following the lead of an autonomous car, traffic congestion could be significantly reduced, wait times minimized, and the overall driving experience streamlined.

A 2021 report by Inrix revealed that poor signal timing is a significant contributor to traffic delays, accounting for approximately 10 percent of total delay time.[55] This figure is double the estimate from seventeen years earlier as stated by the 2004 Federal Highway Administration. However, the 2024 Inrix report states that their own numbers may not reflect the true magnitude of the problem and project that the true delays might be much higher: "Emerging research is concluding that this number is likely much higher, with findings from two states estimating roughly 25 percent of total delay attributed to signals."[56] This data highlights the critical need for improved traffic signal management and the potential benefits of integrating autonomous vehicles into traffic systems. This innovative approach not only boosts traffic efficiency but also heralds a new era of intelligent, interconnected urban transportation systems.

In the upcoming era of smart cities and digital transformation, AI-driven traffic efficiency platforms are emerging as pivotal tools in reshaping urban mobility. These sophisticated systems harness the power of artificial intelligence, machine learning, and big data analytics to analyze and optimize traffic flow, thereby reducing congestion, enhancing road safety, and

---

55 Rick Schuman, "Traffic Signals Meet Big Data," *Inrix* (blog), February 16, 2021, https://inrix.com/blog/suprising-findings-from-the-inrix-signals-scorecard/.

56 Ryan Johnston, "Poorly Timed Traffic Signals Are an Even Bigger Time Waster Than Previously Thought, Report Finds," Nxhut, February 24, 2021, https://nxhut.com/inrix-traffic-signals-traffic-congestion-research-2021/.

improving overall transportation efficiency. By integrating data from a variety of sources, including traffic sensors, cameras, GPS signals, and social media feeds, these platforms provide real-time insights and predictive analytics to traffic management authorities. This enables more informed decision-making, from adjusting traffic signal timings to planning urban infrastructure developments.

The goal is not only to facilitate smoother commutes but also to support sustainable urban planning initiatives by minimizing environmental impact, making AI-driven traffic efficiency platforms a cornerstone of modern urban transportation strategies.

AUTONOMOUS CARS AND MARGINALIZED TRANSPORTATION GROUPS

The emergence of autonomous vehicles heralds a significant shift toward inclusive mobility, offering unprecedented opportunities for those traditionally marginalized from the driving community. This inclusivity primarily benefits the elderly, individuals with disabilities, and those without a driver's license, promising to significantly enhance their quality of life by providing newfound independence and broader access to essential services and opportunities.

For the elderly, autonomous vehicles can mitigate the risks associated with age-related declines in vision, reaction times, and cognitive abilities, allowing them to maintain their independence longer without compromising safety. This technology enables seniors to continue participating in community activities, access healthcare services, and maintain social connections, all of which are vital for their well-being and mental health.

People with disabilities stand to gain immensely from the advent of self-driving cars. These vehicles can be designed with accessible features from the outset, eliminating the need for costly modifications that are often necessary for conventional vehicles. Furthermore, the autonomous nature of these vehicles removes the need for manual control, opening up the possibility of driving for those with physical disabilities who were previously unable to operate a vehicle. Self-driving cars can bring us closer to equal mobility for all, unlocking new levels of independence and making our communities more inclusive than ever.

Moreover, individuals without a driver's license—whether due to age, legal, financial, or personal reasons—will find autonomous vehicles an invaluable asset. This technology offers a reliable and safe mode of transportation, diminishing the reliance on parental or public transit systems, which might not always provide comprehensive coverage or align with the individual's schedule. By ensuring access to personal, autonomous transport, these individuals can more easily access employment opportunities, education, and other essential services, thus enhancing their ability to participate fully in society.

The collective impact of autonomous vehicles on these groups could also extend to the wider community by reducing the demand for public resources, such as paratransit services, and potentially decreasing the overall cost of mobility. By enhancing the efficiency and accessibility of transportation, autonomous vehicles not only promise a future of safer roads but also a more inclusive and equitable mobility landscape, where everyone has the opportunity to move freely and independently.

THE THORNS
INFRASTRUCTURE AND LOGISTICS

Although there are obvious real and theoretical advantages of AI-driven travel, navigating the path to fully autonomous vehicles is fraught with challenges, both anticipated and unforeseen. This will take both time and money—and a lot of thought. The journey toward realizing its full potential is hindered by a complex maze of technological, regulatory, and societal barriers. At the technological level, a significant overhaul of present-day communication technologies and transportation infrastructure is critical. This may include the redesigning of roads to better suit autonomous vehicle navigation as well as necessitating the implementation of clear, consistent road markings and signage. Moreover, the individual ownership model of self-driving cars might paradoxically exacerbate traffic congestion due to the lure of increased convenience. However, a shift toward shared autonomous vehicle services promises to mitigate this by offering an efficient and cost-effective alternative for urban mobility.

Equally crucial is the resolution of regulatory and societal challenges. The integration of autonomous vehicles into the fabric of daily life demands the creation of new legal frameworks tailored to the nuances of autonomous transport. Gaining public trust and acceptance of these technologies, along with ensuring robust protections against cybersecurity risks, stands as a formidable task.

With a steady stream of investment pouring into research and development, however, it's clear there's a dedicated drive to unlock the full potential of autonomous vehicles. This tech evolution is set to completely change how we experience transportation, moving us toward a future defined by greater safety, accessibility, and efficiency.

ETHICS: THE TROLLEY PROBLEM

In the context of autonomous vehicles (AVs), ethical guidelines are critical for navigating complex moral decisions, particularly in scenarios where an unavoidable accident is imminent, and the AV must choose between two harmful outcomes.[57] This dilemma is often referred to as the "trolley problem" in ethics, adapted for the age of autonomous technology.

The trolley problem is a thought experiment in ethics and psychology, illustrating a moral dilemma. First described in 1985 by Judith Jarvis Thomson in the *Yale Law Journal*, it presents a scenario where a trolley is headed toward five people tied up on the tracks.[58] An operator is standing next to a lever that can switch the trolley onto another track, where only one person is tied up. The ethical dilemma for the operator is whether to pull the lever, actively intervening to cause one person's death to save five others, or do nothing and allow the trolley to kill the five people.

The trolley problem raises questions about *utilitarianism* (which suggests the best action is the one that maximizes utility, generally defined as maximizing happiness and reducing suffering) versus *deontological ethics* (which suggests that certain actions are inherently right or wrong, regardless of the outcomes). Pulling the lever would be a utilitarian choice, aiming to minimize overall harm by saving more lives. In contrast, not pulling the lever could be seen as a deontological choice, adhering to a rule that one should not directly cause someone's death.

This dilemma touches on the ethical principles of action

---

57   Patrick Lin, "The Ethics of Autonomous Cars," *The Atlantic*, October 8, 2013, https://www.theatlantic.com/technology/archive/2013/10/the-ethics-of-autonomous-cars/280360/; Thomas H. Davenport, "Getting Real About Autonomous Cars," *MIT Initiative on the Digital Economy* (blog), April 3, 2017, https://ide.mit.edu/insights/getting-real-about-autonomous-cars/.

58   Judith Jarvis Thomson, "The Trolley Problem," *Yale Law Journal* 94, no. 6 (May 1985): 1395–1415, https://doi.org/10.2307/796133.

versus inaction, the value of individual lives, and the moral responsibility of making decisions that affect others. It has been widely discussed in the context of autonomous vehicles, which may face real-life scenarios requiring programmed responses to potential accidents that resemble the trolley problem's structure. But what if the trolley was an autonomous car?

When applied to autonomous cars, the trolley problem becomes a debate on how self-driving vehicles should be programmed to react in situations where an accident is inevitable, and the car must make a choice between two or more harmful outcomes. This brings forth several ethical questions: Should the car prioritize the lives of its passengers over pedestrians? How should it value the lives of young people versus the elderly? Should it attempt to minimize overall harm, even if it means sacrificing its own passengers?

As autonomous vehicles navigate the streets, they may face critical situations requiring instantaneous ethical decisions. Picture an AV driving down a street when suddenly, children dart into the road. The AV is confronted with a split-second and difficult choice. Here are some potential options:

1. Swerve to avoid the children: The AV could swerve, potentially saving the children but risking harm to pedestrians on the sidewalk or other vehicles. This action would prioritize the immediate safety of the children who ostensibly have more life left and potential to contribute to society, but could endanger innocent bystander adults on the sidewalk, not initially involved in the crisis.

2. Maintain course: The AV might also choose to stay its course, avoiding risk to other bystanders but potentially resulting in serious harm to the children. This decision leans toward minimizing broader harm, though at a grave cost.

3. Protecting its autonomous passengers at all costs, regardless of injury to the children and bystanders on the sidewalk.

This dilemma spotlights the intricate ethical decisions AVs must make in milliseconds. The core of the issue revolves around whether to prioritize the number of lives saved, consider the potential future life years of individuals (e.g., children versus older people who have already lived the majority of their lives), or weigh the ethics of actively altering course and potentially endangering its passengers who trusted the machine and who were previously not at risk.

Crafting ethical guidelines for AVs is an intricate process that demands a nuanced approach to these moral quandaries, requiring AVs to make split-second decisions based on complex ethical rules. These guidelines must strike a balance between various factors, such as the potential number of lives affected, the severity of possible injuries, and the ethical implications of action versus inaction.

While there may not be a universally correct choice in such scenarios, they underscore the challenge of embedding AI within contexts where ethical and moral decisions are paramount. It underscores the need for comprehensive ethical frameworks, crafted collaboratively by technologists, ethicists, legal experts, and the public, to guide AV behavior in critical situations.

The ethical programming of AVs brings to light broader considerations, including liability, the legal frameworks governing AV operations, and the transparency of the decision-making algorithms. The programming of these vehicles reflects ethical decisions that must be debated and agreed upon broadly to foster public trust and acceptance of this transformative technology. The quest for consensus on these ethical guidelines is challenging, given the wide variation in moral intuitions and principles among individuals.

Navigating these ethical dilemmas is not just a technological challenge but a societal one, highlighting the profound implications of integrating advanced AI into our daily lives and the roads we travel.

INTELLECTUAL PROPERTY VERSUS TRANSPARENCY AND ACCOUNTABILITY

In the midst of advancing autonomous vehicle (AV) technology, the trolley problem reemerges not just as a philosophical debate but as a real-world dilemma that these vehicles may face. This issue intensifies the call for transparency and accountability in the decision-making processes of AVs, particularly when navigating the murky waters of intellectual property.

The heart of the matter lies in the decision-making algorithms of AVs—complex, proprietary systems that are often shielded from public and regulatory scrutiny. This opacity raises significant concerns: When an AV makes a critical decision, how can the public be assured that it was the right one? And when decisions lead to accidents, determining accountability becomes a convoluted issue. Unlike traditional vehicles, where the driver's responsibility is clear, the responsibility for AVs' actions could fall on various shoulders: the manufacturer, the software developers, or even the entities that maintain the vehicle's operational environment.

To untangle these ethical knots, there's a pressing need for enhanced transparency. This means pushing for standards that make AI's decision-making processes accessible and understandable, not just to experts but to the public at large. However, this drive for openness clashes with the desire of companies to protect their intellectual property. The algorithms that guide AVs are not just lines of code but valuable assets, the product of substantial investment in research and development. Companies naturally seek to protect this intellectual property, creating a tension between commercial interests and the public's demand for transparency.

Addressing this tension requires a multifaceted approach. On one hand, ethical decision-making frameworks could be integrated into AV systems, developed through broad societal input and made publicly available to build trust. On the other hand, regulatory bodies might mandate the recording and storage of detailed decision-making data in AVs, akin to an aircraft's black box, which could be analyzed after any incident to determine accountability.

The development of shared accountability models is another avenue to explore, distributing responsibility among all stakeholders involved, from manufacturers to software developers, and even including regulatory bodies. These models, alongside legal frameworks, should offer clear paths for recourse for those affected by AV decisions, ensuring that the victims of accidents are not left without support due to the faceless nature of machine decision-making.

CLOSING THOUGHTS: GUIDING THE ROAD AHEAD WITH PURPOSE

As we move further into the realm of autonomous vehicles, we're reminded of the transformative impact that Carl and Bertha Benz's invention had on society—ushering in a new era of mobility. Today's advancements in autonomous technology hold a similarly revolutionary promise, but the road forward is laden with complex challenges. From ethical dilemmas and regulatory concerns to technical and social issues, our path forward requires not only innovation but also a deep commitment to responsibility.

To navigate these challenges, developing and implementing ethical AI frameworks is essential. These frameworks must be grounded in justice, equality, and respect for human rights, ensuring that AI systems prioritize fairness and inclusivity. Engaging a broad range of stakeholders—including ethicists, sociologists, technologists, and impacted communities—is crucial to identifying and addressing biases, bringing transparency and accountability into the development process. Such interdisciplinary collaboration encourages the creation of AI systems that respect privacy, promote social good, and honor the values of the public they're designed to serve.

By embracing transparency, fostering ethical frameworks, and balancing innovation with responsibility, we can harness the potential of autonomous vehicles and AI-driven technologies while addressing the profound ethical and societal challenges they introduce. This approach ensures that our journey toward a more automated future is not only technologically advanced but also guided by principles that protect the well-being and rights of all individuals.

Just as the Benzes' first drive turned roads into paths of possibility, our work today could pave the way to a safer, more inclusive transportation landscape—a vision worth pursuing with both optimism and caution.

CHAPTER 6

THE COST OF RECHARGEABLE BATTERIES

"Let him who is without sin among you throw the first stone."

—JOHN 8:7

I fell in love with endurance sports in my early twenties.

This passion has since led me to complete more than fifty-seven marathons, including twice finishing the Badwater 135 ultramarathon across Death Valley to the top of Mount Whitney, often touted as the toughest human endurance footrace in the world. Over time, my pursuits expanded to include ultra-distance cross-country skiing, Ironman competitions, and ultra-cycling events. Each new challenge fueled my desire to push both my physical and mental limits, constantly testing how far I could go before my own "batteries" needed recharging.

For my fifty-fifth birthday, I decided to tackle Pete Stetina's Paydirt—a grueling seventy-mile gravel bike race across the Pine Nut Mountains, set against the snowcapped Sierra Mountains surrounding Carson Valley, Nevada. The race, like so many today, was steeped in technology, with riders decked out in heart monitors, power meters, and cadence trackers. The winner is determined by competing across three timed Strava segments, all meticulously mapped and recorded via smartphone.

As I stood at the start line that cool morning, I wondered if I still had the "charge" to keep up. I couldn't help but reminisce about my first bike race more than forty years before—when things were simpler. Back then, you just started and finished. Now, tech was everywhere, as integral to the race as the riders themselves—maybe even more so.

The night before, I camped at Fuji Park, surrounded by hundreds of Mercedes-Benz Sprinter vans, each outfitted with Starlink internet and a host of rechargeable batteries, maintaining the conveniences of home. It was a far cry from the simple pop-up tents I remembered from my early days in the sport. In my forty years as a cyclist, I have witnessed the riding community's culture evolve from camping in simple pop-up tents to living out of these behemoth battery-packed mobile homes.

The morning of the race, the parking lot filled with Teslas—evidence of the tech-driven wealth that now dominates the Tahoe and Northern Nevada regions. The bikes, too, had changed. Every mountain or gravel bike came equipped with Di2 battery-powered shifters—precision technology that carried a price tag of $8,000–$12,000.

But the bike racers of today—myself included—don't just have one bike; we typically have three: a gravel bike, a mountain bike, and a road bike—all featuring this costly battery technology.

Rechargeable batteries have become essential to our modern lives, powering everything from our phones to our bikes. Smartphones have become indispensable tools, enabling seamless communication, entertainment, and productivity. Similarly, tablets and e-readers let us carry entire libraries, allowing us to read or browse without being tethered to a power outlet.

Wearable technology, like smartwatches and fitness trackers, monitor our health, track fitness activities, and keep us connected, while wireless headphones and earbuds free us from the hassle of cords, offering high-quality sound and mobility for hours on end.

Electric vehicles are revolutionizing transportation by providing a sustainable alternative to gasoline-powered cars. In-home improvement and construction power tools such as cordless drills and saws deliver flexibility and efficiency, making tasks easier to manage.

In the medical field, devices like pacemakers, hearing aids, and portable diagnostic equipment enhance our health and longevity. At home, cordless vacuum cleaners, lawnmowers, and kitchen gadgets make daily chores simpler and more efficient, offering the freedom to move about without the limitation of power cords.

On race day, these rechargeable batteries were everywhere—in the bikes, the Strava apps, the heart rate monitors, the Teslas, and the Sprinter vans. They had seamlessly woven themselves into the fabric of our lives, powering both work and play. In the journey from stones to phones, we've witnessed the incredible progression of technology and its profound impact on society.

But how often do we stop to ask where the minerals that make these batteries come from? What's the hidden human cost of the tech we rely on? How do batteries even work?

I will answer these questions, and more, in this chapter.

As we embrace the future of EV technology, we see it as a giant leap toward achieving net-zero carbon emissions. This progress builds on the ingenuity of countless innovators, from the creators of horse-drawn carriages to the pioneers of the combustion engine.

AI promises a greener future by accelerating the adoption of electric vehicles and renewable energy storage. But here's the paradox: The very batteries that power this progress rely on lithium and cobalt—minerals whose extraction damages ecosystems and exploits vulnerable workers. Can we reconcile these competing demands for a truly sustainable future? In this chapter, we uncover the hidden costs behind our rechargeable lifestyles by exploring the extraction processes for lithium and cobalt—the core elements powering the devices, vehicles, and conveniences of modern life. Drawing on insights from Siddharth Kara's *Cobalt Red*, we illuminate the pressing need for a more responsible approach to battery technology. His groundbreaking research serves as the foundation for much of this discussion, exposing the environmental and human toll behind the minerals that fuel our digital and AI-driven world.

The Paradox: AI accelerates demand for battery-powered devices, but this reliance on lithium and cobalt extraction exacerbates environmental and human rights issues.

The Roses: AI-driven innovation in EVs and energy storage reduces carbon emissions.

The Thorns: Mining for materials damages ecosystems, exploits workers, and creates supply chain vulnerabilities.

But first, how do batteries even work? Because I like rechargeable batteries. And you like rechargeable batteries.

We *all* like rechargeable batteries.

HOW BATTERIES WORK

Immediately after the Paydirt race, I took a bar of soap, housed in a ziplock bag, hosed off in the park, changed clothes, and headed to the Reno airport to catch a red-eye flight to Boston. I was on my way to the MIT campus for a week of intense learning at the Future Compute and EM Digital courses on emerging artificial intelligence platforms.

From satellite dishes to high-tech vans, from advanced bike races to smartphones and tablets, and jet-setting around the globe, those of us in the first world enjoy an era of unprecedented convenience and technological marvels. Modern life in the first world is truly astonishing, offering ease and efficiency like never before in history.

But pause for a moment: Is it like this for *everyone*?

By the next morning, I found myself in a completely different culture than that of the dusty Western mountain bike race in which I had been just twelve hours before. Waiting among the intelligentsia and lost in the bustle of Cambridge, Massachusetts, the Co-Op bookstore at Harvard Square was about to open.

I had one thing on my mind: I wanted to get my hands on a book I'd been meaning to read. There it was, stacked quietly on a shelf under "Government and Politics"—*Cobalt Red: How the Blood of the Congo Powers Our Lives* by Siddharth Kara.

I rarely get time to just sit in a café with a book—that's something *other* people do. But that morning, in the crisp air of a perfect spring day, I found myself sitting at a Harvard Square café, sipping black coffee and reading about the dark side of the convenience we all enjoy.

The book—which formed the genesis of this chapter—made me pause. Do any of us even know how batteries are made?

Let's break down how a battery works. A battery is like a

container that stores electrical energy for use when needed. Inside a battery, there are three main parts: the anode (negative side), the cathode (positive side), and the electrolyte (a special chemical that allows electricity to flow between the anode and the cathode). Think of it like a water park ride with two big water tanks: one at the top (the anode) and one at the bottom (the cathode). The water (electrons) wants to flow from the top tank (anode) to the bottom tank (cathode) because water naturally flows downhill. However, there's a barrier (electrolyte) between the two tanks, so the water can't flow directly. To make the water flow, we need a special path, like a water slide, that connects the two tanks. When we connect a device (like a flashlight or a phone) to the battery, it's like opening the slide, allowing the water (electrons) to flow from the top tank (anode) to the bottom tank (cathode) through the device, which uses the energy to work.

Here's what happens in more detail: Electrons build up in the anode (negative side). This anode is usually made of a carbon-based structure like graphite. When you connect a device to the battery, a chemical reaction occurs in the anode, releasing electrons. These electrons flow through the external circuit (your device) to get to the cathode (positive side), creating an electric current that powers your device. The battery works by moving electrons from the anode to the cathode through your device, providing it with power.

When the battery "runs out," it means most of the electrons have moved from the anode to the cathode, and the battery needs to be recharged to reset this process. Recharging a battery is like pumping the water back up to the top tank so it can flow down again, done by reversing the chemical reactions using an external power source (like plugging your phone into a charger).

Lithium-ion batteries rule the game because lithium is light, which makes it ideal for the tech we carry everywhere. But here's where cobalt comes in. Cobalt is used in the cathodes due to its unique electron-deficient configuration. It stabilizes the battery, prevents overheating, and lets it run at high energy densities—meaning more charge, longer battery life, and fewer fires or explosions. Cobalt is what makes your battery last longer and perform better, especially through countless charges and discharges. It's a crucial player in the battery world, especially for devices that need to be compact yet powerful.

So, where does it come from?

Imagine, if you will, a far-off exotic land blessed with riches in natural resources beyond the scope of human comprehension—ivory, palm oil, copper, tin, zinc, silver, gold, nickel, diamonds, uranium, tantalum, tungsten, and most importantly, cobalt. This silver-blue metal is fueling a frantic, almost frenzied rush to extract resources at any cost. In the Congo, men, women, and even children are still mining cobalt the old-fashioned way—picks, shovels, and bare hands—enduring brutal, dangerous conditions.

Why do they do this? So we can have rechargeable bikes, smartphones, and electric vehicles.

The kicker? While researchers scramble to find alternatives to cobalt, nothing yet matches its ability to keep batteries stable and efficient. It's why lithium-ion technology still relies on cobalt to keep our tech humming along.

But at what cost?

As I sat there, the weight of it all hit me. Our world is powered by minerals mined in ways that are far from magical. And it begs the question: What are we willing to overlook for the sake of convenience? Is there a way forward where we can enjoy the benefits of modern technology *without* sacrificing the well-being of others or the planet?

That's the conversation we need to start having—because the demand is only going to increase.

THE THORNS
POWER SURGE: THE RISING DEMAND FOR BATTERIES

The lithium-ion battery market experienced its first significant surge in demand with the smartphone and tablet revolutions. It all started when Apple introduced the iPhone in 2007, followed by the launch of Android in 2008. By 2010, Apple released the iPad, and soon Samsung followed suit with its Galaxy line. Since then, billions—yes, billions—of smartphones and tablets have been sold. And that's not even counting other battery-hungry devices like e-bikes, electric scooters, earbuds, and wearable fitness technology, all of which also rely on lithium-ion technology.

But it was the Paris Agreement of 2015 that truly ignited the electric vehicle market. When 195 nations committed to limiting the global temperature rise to below 2 degrees Celsius from preindustrial levels, a massive shift began.[59] To reach this goal, CO_2 emissions must drop at least 43 percent by 2030 compared to 2019 levels.[60] With roughly 28 percent of all CO_2 emissions coming from traditional gas-powered vehicles, the push to electrify transportation kicked into high gear.[61]

Meanwhile, the EV30@30 initiative set an ambitious target: by 2030, 30 percent of all global vehicle sales should be electric,

---

59 FCCC. CP. 2015/L.9, Adoption of the Paris Agreement (Dec. 12, 2015), https://docs.un.org/en/fccc/cp/2015/l.9.

60 "For a Livable Climate: Net-Zero Commitments Must Be Backed by Credible Action," United Nations, accessed March 20, 2025, https://www.un.org/en/climatechange/net-zero-coalition.

61 "Fast Facts on Transportation Greenhouse Gas Emissions," United States Environmental Protection Agency, last updated June 18, 2024, https://www.epa.gov/greenvehicles/fast-facts-transportation-greenhouse-gas-emissions.

amounting to around 230 million electric vehicles.[62] For context, in 2010, there were just 17,000 electric vehicles worldwide.[63] The momentum grew even stronger at COP26, where thirty nations pledged to phase out gas-powered vehicles by 2040, reinforcing the urgent need for long-lasting, high-performance batteries.[64]

This rapid growth in smartphones, wearables, and EVs is driving an insatiable demand for lithium and cobalt. Between 2017 and 2025, the demand for lithium is projected to more than triple, from 236,000 tons (214 kilotons) to a staggering 737,000 tons (669 kilotons) of lithium carbonate equivalent (LCE). Cobalt isn't far behind, with demand expected to jump 60 percent, from 150,000 tons (136 kilotons) to 245,000 tons (222 kilotons) of refined metal equivalent. These numbers assume that lithium-ion technology will continue to dominate the battery market, according to McKinsey & Company.[65]

While these changes are essential for societal progress—driving innovation in smartphones, tablets, and other technologies that enhance human capabilities—and for reducing carbon emissions to achieve a net-negative footprint compared to combustion engines, they come with significant environmental and human costs. The millions of tons of lithium and cobalt

62 Timur Gül et al., *Global EV Outlook 2021: Accelerating Ambitions Despite the Pandemic* (International Energy Agency, 2021), 13, 73, https://iea.blob.core.windows.net/assets/ed5f4484-f556-4110-8c5c-4ede8bcba637/GlobalEVOutlook2021.pdf.

63 Timur Gül et al., *Global EV Outlook 2020: Entering the Decade of Electric Drive?* (International Energy Agency, 2020), 11, 39, https://iea.blob.core.windows.net/assets/af46e012-18c2-44d6-becd-bad21fa844fd/Global_EV_Outlook_2020.pdf.

64 "COP26: Together for Our Planet," United Nations, accessed March 20, 2025, https://www.un.org/en/climatechange/cop26.

65 Marcelo Azevedo et al., *Lithium and Cobalt—A Tale of Two Commodities* (McKinsey & Company, June 2018), 8, https://www.mckinsey.com/~/media/mckinsey/industries/metals%20and%20mining/our%20insights/lithium%20and%20cobalt%20a%20tale%20of%20two%20commodities/lithium-and-cobalt-a-tale-of-two-commodities.pdf.

required for mining create substantial impacts. Although electric vehicles have a significantly lower cumulative life cycle emission than traditional cars, it's crucial not to overlook or dismiss the consequences they still carry.

Unbiased knowledge is key to real progress. This situation highlights the complex, nonlinear nature of advancement. We must balance our excitement for a greener future with the realities of resource extraction. It should prompt a thoughtful dialogue on improving battery technology, reducing the environmental toll of mining, exploring alternative mineral and energy sources, and—most importantly—enhancing the conditions for the people extracting these critical resources.

So, what are the true costs of the rechargeable battery revolution? And as users of this technology, what should we understand to push for a more informed, responsible path forward?

WHAT BATTERY USERS NEED TO KNOW

In order for us to make the appropriate, thoughtful decisions regarding the future of batteries, there are three areas battery users need to know: the weight and energy density of EV batteries and the massive earthmoving required to produce them; the environmental impacts of lithium extraction, particularly its strain on water resources; and the ethical concerns surrounding cobalt mining, where human exploitation and dangerous conditions persist.

By understanding these challenges, we can better advocate for solutions that balance technological progress with sustainability and humanity. Let's go over each one.

Weight, Energy Density, and Machine Earthmoving

The weight of an EV battery varies depending on the vehicle and model, but the average is 1,000 pounds (454 kg). In some models, however, it can reach up to 2,000 pounds (900 kg). The heavier the battery, the more energy it can store, although this relationship is not always straightforward. While larger batteries can increase a vehicle's range, the added weight also requires more energy to power the vehicle.

A crucial factor in EV battery performance is energy density, which refers to the amount of energy stored per unit of weight. Higher energy density allows for more energy storage without increasing the battery's weight, which is essential for maximizing an EV's range. Lithium, the lightest metal on the periodic table, is crucial for achieving high energy density. Lithium-ion batteries, with an energy density of about 260–270 watt-hours per kilogram (W/kg), are preferred for applications requiring long battery life, such as phones, laptops, and EVs. In contrast, traditional lead-acid batteries offer only 50–100 W/kg.[66]

However, the production of EV batteries is highly resource- and capital-intensive. The raw materials required for these batteries are typically mined using carbon-emitting, combustible-engine machinery or human labor, which we will discuss later. A typical lithium-ion battery for an EV contains around thirty pounds of cobalt, though this amount can vary depending on the battery's size.[67] For example, a Tesla Model S battery contains approximately 138 pounds of lithium, although the specific amount of cobalt it holds is not disclosed.[68] In addi-

---

66 "Electric Car Battery Weight Explained," *EVBox* (blog), May 4, 2023, https://blog.evbox.com/ev-battery-weight.

67 Teague Egan, "Where Do Electric Vehicle Batteries Come From?," EnergyX, March 10, 2023, https://energyx.com/blog/where-do-electric-vehicle-ev-batteries-come-from/.

68 *EVBox* (blog), "Electric Car Battery Weight."

tion to cobalt and lithium, EV batteries contain other metals such as manganese (around forty-five pounds) and nickel (about 110 pounds). Lithium-ion batteries are also composed of other materials, including copper, aluminum, and graphite. By weight percentage, a typical lithium-ion battery consists of 7 percent cobalt, 7 percent lithium, 4 percent nickel, 5 percent manganese, 10 percent copper, 15 percent aluminum, 16 percent graphite, and 36 percent other materials.

The production of each EV battery requires the processing of vast quantities of raw materials: approximately 25,000 pounds of brine for lithium, 30,000 pounds of ore for cobalt, 6,000 pounds of ore for nickel, and 25,000 pounds of ore for copper. This amounts to the extraction of 500,000 pounds of the earth's crust for just one battery.[69]

That's a staggering amount of raw materials, underscoring the massive mineral demands and the heavy environmental toll of producing just one EV battery.

Lithium and Water

Lithium might be the backbone of your favorite gadgets and EVs, but getting it out of the ground is no small feat. In fact, it's a water-guzzling process that leaves a serious mark on some of the world's driest places.

More than 95 percent of the world's lithium supply comes from brines of lithium-rich hard-rock ores, with significant production occurring in Australia, China, and Chilean Latin America. In Chile's Atacama Desert—one of the driest places on earth—you'll find hundreds of massive turquoise pools, some

---

69 Mark P. Mills, "Mines, Minerals, and "Green" Energy: A Reality Check," Manhattan Institute, July 9, 2020, https://perma.cc/UE9K-KQKN.

as big as twenty football fields. These pools are filled with saline brine drawn from ancient underground reservoirs, rich in lithium carbonate, which can be processed into lithium.[70]

In a desert already suffering from extreme water scarcity, lithium companies pump billions of liters of brine from these reservoirs. Chilean company SQM and US-based Albemarle extract nearly 528 gallons (2,000 liters) per second—over 16.64 billion gallons (63 billion liters) a year. The water-intensive process evaporates up to 95 percent of the water using solar radiation, heavily impacting local water resources. And while this brine isn't drinkable, its removal still disrupts the Atacama's delicate water cycle.

The loss of brine exacerbates water shortages, creating major challenges for the region's residents. Estimates on water use for lithium extraction range from 106 gallons to a staggering 528,344 gallons (400 to 2 million liters) per kilogram of lithium. While Albemarle claims it's on the lower end, they've never revealed how they calculated that figure. There's real concern that brine extraction is mixing with freshwater resources, increasing salinity and making the water undrinkable for local communities.

Cobalt and Human Exploitation

While lithium is primarily mined using heavy machinery, cobalt extraction is a different story.

Ironically, the development of lithium-ion technology using lithium cobalt oxide (LCO) was first introduced by the fossil fuel giant Exxon during the OPEC oil embargo of the 1970s.

70 Nikolaj Houmann Mortensen i Mie Obbekær, "How Much Water Is Used to Make the World's Batteries?," Electronica Justa, accessed March 20, 2025, https://electronicajusta.net/portfolio/ quanta-aigua-sutilitza-per-fabricar-les-bateries-del-mon/?lang=en.

At that time, the term "alternative energy source" surfaced as a financial safeguard in case oil supplies became difficult to secure. Cobalt, aside from its role in batteries, is also used to cleanse fossil fuels in the production of refined gasoline. Additionally, its resistance to wear makes it ideal for manufacturing superalloys used in turbine and jet engines, toughening tools, and even components in artificial joint replacements.

According to author Siddharth Kara, in his book *Cobalt Red*, the EU has designated cobalt as one of twenty "critical" metals of the world, while the United States and China have both designated cobalt as a highly "strategic" material. As the tech and arms race between these superpowers intensifies, reducing China's dominance in global cobalt mining has become a key strategic goal for both the US and the EU.

Due to geological processes that occurred millions of years ago, deposits of cobalt are unusually shallow and often lie just a few feet below the surface. According to the US Geological Survey in 2022, the Central African Copper Belt holds more than half of the world's cobalt reserves, estimated at 3.5 million tons.[71] Most of this cobalt is located within the Democratic Republic of Congo at very shallow depths, making it accessible by hand and shovel.

Kara highlights the grim reality behind the global cobalt supply chain, where the most advanced electronics rely on cobalt extracted by impoverished workers using simple tools:

> The most advanced consumer electronics devices in the world rely on a substance that is excavated by the blistered hands of peasants using picks, shovels, and rebar. Labor is valued by the penny, life hardly at all. The global cobalt supply is the mechanism that

71   U.S. Geological Survey, *Mineral Commodity Summaries 2022* (U.S. Geological Survey, 2022), 53, https://doi.org/10.3133/mcs2022.

transforms the dollar-a-day wages of the Congo's artisanal miners into multibillion-dollar quarterly profits at the top of the chain...

[Thousands] of women, children and men shoveled, scraped and scrounged across the artisanal mining zone under a ferocious sun and a haze of dust...

Samples of the dust taken inside the homes throughout the Copper Belt had an average of 170 micrograms of lead per square foot... By way of comparison, the Environmental Protection Agency in the United States recommends a maximum safe limit of 40 micrograms of lead per square foot inside homes. Levels as high as 170 micrograms per square foot can cause neurological damage, muscle and joint pain, headaches, gastrointestinal ailments, and reduced fertility in adults. In children lead poisoning can cause irreversible developmental damage as well as weight loss, vomiting, and seizures...

In the studies we conducted, the artisanal miners have more than forty times the amount of cobalt in their urine as the control groups. They also have five times the level of lead and four times the level of uranium...

Inhalation of cobalt dust causes "hard metal lung disease" which can be fatal...

Cancers were also on the rise in artisanal mining communities, especially of the breast, kidney, and lung...

According to the OECD (Organization for Economic Co-operation and Development), up to seventy percent of the cobalt from the DR Congo has some touch with child labor...

Foreign mining companies would argue that they do not employ artisanal miners, so the responsibility is not theirs, even though the cobalt from artisanal digging ends up in their supply chains... The government of the DRC would argue that they do not have the money to support good wages or other income schemes, even though the mining concessions are sold for billions of dollars and royalties and taxes in the billions are collected each year based in no small part on the value of the minerals excavated by artisanal miners. Cobalt refineries, battery manufacturers, and tech and EV companies would argue that the responsibility should be borne downstream, even though the scramble for cobalt only exists because of their demand for it. Therein lies the great tragedy of the Congo's mining provinces—no one up the chain considers themselves responsible for the artisanal miners, even though they all profit from them.[72]

But this isn't the first time that the Congo has supplied natural riches to the world's capitalistic empires. In the late 1800s, the region provided ivory for piano keys and false teeth. By the 1890s, it was rubber for car and bicycle tires. In the early 1900s, palm oil became a major export for soap production. The 1910s saw the Congo supplying tin, copper, zinc, silver, and nickel for industrial alloys. Throughout history, it has consistently provided diamonds and gold to global luxury markets. During World War I, millions of bullets fired by British and American forces were made from Congolese copper.[73] By 1945, uranium from the Congo was used in the nuclear bombs of World War II. In the 1990s and early 2000s, tantalum and tungsten were

---

72   Siddharth Kara, *Cobalt Red: How the Blood of the Congo Powers Our Lives* (St. Martin's Press, 2023), 13, 24, 52, 61–62, 65, 67.

73   Dan Snow, "DR Congo: Cursed by Its Natural Wealth," BBC, October 9, 2013, https://www.bbc.com/news/magazine-24396390.

extracted for microprocessors. And finally, since around 2012, cobalt mining for rechargeable batteries has become the latest chapter in this long history of resource exploitation.

According to Darton Commodities—a global commodities trading company that specializes in the supply and marketing of cobalt and other critical metals—global cobalt production grew by 23 percent in 2022, eliminating the large deficit seen in 2021. This surge was primarily driven by increases in production from the Democratic Republic of Congo (DRC), which now accounts for about 75 percent of global supply, and Indonesia, which is emerging as a major player in the sector. Darton also reported that global mine production increased by 42 percent between 2020 and 2022, driven by the easing of COVID-19-related supply chain constraints, expansion at existing operations, and the opening of several new mines. In 2022, Glencore became the world's largest cobalt miner, primarily due to its operations in the Congo.[74]

In the Democratic Republic of Congo (DRC), major mining companies exploit human labor instead of relying on machinery for several key reasons, all motivated by profit at the expense of workers' well-being. First and foremost, labor—often child labor—is significantly cheaper than importing and maintaining advanced machinery. The socioeconomic conditions in the DRC contribute to this exploitation. According to the United Nations Human Development Index—a composite measure that ranks countries based on life expectancy, education, and income—the DRC ranks 180th out of 193 countries.[75] Child

---

74 Mark Burton and Michael J. Kavanagh, "The Cobalt Market Saw a Record-Breaking Supply Boom in 2022," *Bloomberg*, March 7, 2023, https://www.bloomberg.com/news/articles/2023-03-06/the-cobalt-market-saw-a-record-breaking-supply-boom-in-2022.

75 United Nations Development Programme, "Human Development Index (HDI)," United Nations, accessed April 23, 2025, https://hdr.undp.org/data-center/human-development-index#/indicies/HDI.

mortality in the DRC is the eleventh highest in the world, access to clean drinking water is only 59.4 percent, and electrification is at a mere 21.5 percent.[76]

Additionally, the World Bank reported that as of 2022, the adult literacy rate for women aged fifteen and older in the DRC was 71.7 percent, with most workers earning only $1 to $2 per day.[77] These harsh realities make cheap labor abundant and attractive for mining companies eager to maximize profits. The DRC's underdeveloped infrastructure presents additional challenges for mechanization. Poor road conditions, unreliable electricity, and limited access to spare parts further foster the narrative that it is difficult to transport and maintain heavy machinery. This infrastructure deficit makes it convenient for mining companies to exploit human labor than to invest in heavy machinery.

As a result, artisanal and small-scale mining (mining done by hand) plays a significant role in cobalt extraction in the DRC. Although artisanal mining provides employment for many local communities, the conditions are often dire, with workers facing significant hazards. Flexibility and accessibility also favor manual labor. Human miners can access small, irregular, or hard-to-reach ore deposits that large machinery would struggle with. This is particularly relevant in areas where cobalt is found in narrow veins or mixed with other minerals. Moreover, the lack of regulatory oversight and safety standards in the DRC allows for the continued use of child labor despite the hazardous conditions. This regulatory void benefits mining

---

76 "Explore All Countries—Congo, Democratic Republic of the," The World Factbook, last updated April 17, 2025, https://www.cia.gov/the-world-factbook/countries/congo-democratic-republic-of-the/.

77 "Literacy Rate (%)," World Bank Group Gender Data Portal, accessed March 20, 2025, https://genderdata.worldbank.org/en/indicator/se-adt; "The World Bank in DRC: Overview," World Bank Group, last updated October 18, 2024, https://www.worldbank.org/en/country/drc/overview.

companies, enabling them to sidestep the costs associated with ensuring worker safety. The exploitation of cheap labor and the absence of proper regulations create a perilous environment for miners, who face biohazards, toxic exposure, and numerous dangers within the mines.

The book *Cobalt Red* highlights the importance of global strategic moves and countermoves among superpowers in controlling cobalt production and manipulating cheap labor. In 2006, during the Sino–African summit in Beijing, China cemented its control (monopoly, really) over cobalt mining in the DRC. A deal was struck between SICOMINES (short for Sino–Congolese mining) and President Joseph Kabila, where SICOMINES (through a series of shady deals and multimillion-dollar kickbacks to Kabila's family) agreed to invest $6 billion toward the construction of roads and an additional $3 billion toward mining infrastructure upgrades. In exchange, China would be repaid through the value of copper and cobalt extracted from the DRC. The caveat, heavily favoring the Chinese, stipulated that if the mines did not produce as expected, the DRC would repay the loans through "other means."

In 2008, under pressure from the IMF and World Bank, Kabila renegotiated the terms of the agreement. The "other means" clause and the additional $3 billion obligation were dismissed. Under the new terms, SICOMINES agreed to pave 4,101 miles (6,600 km) of roads, build several hospitals, and construct two universities.[78]

Further solidifying this power play, the production of EV batteries is concentrated within a few companies, predomi-

78   Johanna Jansson et al., *Chinese Companies in the Extractive Industries of Gabon & the DRC: Perceptions of Transparency* (Centre for Chinese Studies, University of Stellenbosch, 2009), 33–34, https:// web.archive.org/web/20110726212315/http://www.ccs.org.za/wp-content/uploads/2009/11/Chinese_Companies_in_the_Extractive_Industries_of_Gabon_and_the_DRC._CCS_report_August_2009.pdf.

nantly in China, which holds 70 percent of the production capacity for cathodes and 85 percent for anodes. Contemporary Amperex Technology Co. Limited (CATL) leads the global market with a 34 percent share, largely due to China's control over more than half of the raw materials needed for battery production. Following CATL, LG Energy Solution from South Korea holds a 14 percent market share and has recently partnered with Honda to build a $4.4 billion EV battery plant in the United States, set to begin production in 2025. The third-largest manufacturer, BYD, also based in China, controls 12 percent of the market and uniquely meets most of its own battery and EV system needs due to its dual role as an EV manufacturer.[79]

Other key players in the global EV battery market include Japan and South Korea, which account for 11 percent and 14 percent of production, respectively, while the United States holds a 7 percent share. Despite efforts from the US and EU to boost domestic production through public sector initiatives, China is expected to maintain its dominant position in EV battery manufacturing until at least 2030.[80]

Global ownership and control over cobalt mines illustrate the roses and thorns of progress in EV battery production. While the production of battery-powered vehicles generates more carbon dioxide than manufacturing gasoline-powered cars, EVs ultimately prove to be more environmentally friendly. A recent March 2024 report in *Scientific American* highlights that an average EV produced in the US in 2023 will overcome the initial emissions disparity in approximately 2.2 years or 25,000 miles. After this point, EVs emit significantly less CO_2

---

79 "What Are Electric Car Batteries Made Of?," *EVBox Blog*, August 3, 2023, https://blog.evbox.com/uk-en/what-are-ev-batteries-made-of.

80 *EVBox Blog*, "Electric Car Batteries."

compared to internal combustion vehicles, particularly considering the typical lifespan of a car. Despite the imperfections of EV technology, it ends up being a cleaner alternative in terms of CO_2 emissions.[81]

While the advancement of electric vehicles offers net environmental benefits, such as reduced carbon emissions, and represents a leap forward toward the goal of net-zero emissions, it also comes with substantial material, environmental, and human costs—probably more than most rechargeable batteries users realize. Like all technology at its inception, it is not perfect. China's dominance in this sector further complicates the landscape, presenting strategic advantages, ethical dilemmas, and the potential for global conflict over limited resources.

This paradox of progress underscores the need for unbiased knowledge as the first step toward fostering an informed dialogue. Every rechargeable battery user must understand these issues to contribute meaningfully to discussions on how we can improve and progress as a society.

THE ROSES
A NECESSARY TRANSITION

Lithium-ion batteries represent a critical step toward a cleaner, more sustainable future. They've enabled the rise of electric vehicles, reduced our reliance on fossil fuels, and accelerated the shift toward renewable energy. But as with any technological advancement, they come with trade-offs. While they offer a path away from carbon-intensive energy sources, they also

---

81   Mike Lee, "Electric Vehicles Beat Gas Cars on Climate Emissions over Time,"
      Scientific American, March 12, 2024, https://www.scientificamerican.com/article/
      electric-vehicles-beat-gas-cars-on-climate-emissions-over-time/.

introduce new challenges—environmental degradation, supply chain exploitation, and geopolitical tensions over rare minerals.

That leaves us with an uncomfortable question: When it comes to lithium-ion battery technology and the untold environmental and human costs that go along with it, who is to blame?

Do we point the finger at the Democratic Republic of Congo for corruption and human exploitation? Or at China for being the next of many colonial empires monopolizing cheap mineral resources? Should we blame the lucrative mining companies or the multibillionaire tech moguls who have maximized profits while knowing all along where the cobalt and lithium in their supply chains come from?

Or should we point the finger at ourselves, coasting along on rechargeable bikes, wearing rechargeable watches, glued to our smartphones and tablets, and now turning to emerging EV cars as if it's a clean break?

The truth is, we're all in the hot seat. Since we are all responsible in some way, perhaps we shouldn't cast stones at anyone.

Rather, we need to realize that significant chapters in our history are yet to be written, particularly regarding sustainability and our ability to power the world's ever-increasing population, technology, and energy demands. Several inflection points still lie ahead. The interesting thing is that inflection points, especially those that exist between utopian and dystopian views, attract the most attention. And, if nothing else, attention stimulates dialogue, debate, and innovation.

Which path will we collectively choose? Which conversation will we have? Will it be healthy, respectful, or adverse?

One choice offers a well-known path: continuing to burn fossil fuels, a limited commodity with known environmental costs, until their exhaustion—and possibly ours. Another choice is to pursue EV technology as we know it and go no further, with lithium and cobalt at the forefront, risking continued environmental destruction with marginal gains over fossil fuel energy while ignoring known humanitarian, environmental, and potentially large geopolitical conflicts (war) associated with these limited commodities.

Haven't we been here before?

There's also a third option: redefining how we think about lithium-ion batteries. Although they are the de facto rechargeable batteries of today, what if we all—fossil fuel and EV lovers alike—come to the middle and view them as a "transitional technology"? A stepping stone to something much better? Is it possible, now that we have proven that lithium-ion EV technology is slightly better than fossil fuels, to capitalize on emerging technology like AI to find alternative mineral sources or develop new materials that may not yet exist? What would this alternative reality look like?

If we view both combustion engines and lithium-ion batteries—with all their problems—as mere bridges toward something greater, we open the door to a paradigm shift. It's a choice between embracing innovative solutions that truly support a sustainable future, or continuing on a path that compromises both the planet and our collective well-being.

After all, we've done it before.

TRANSITIONAL TECHNOLOGIES

Throughout history, technological progress has rarely arrived in a single leap; instead, each advancement has built upon those that came before it, driving us step-by-step toward greater achievements.

Take transportation, for example. From the wheel to the

horse and cart, to the combustible engine, and now to electric vehicles powered by lithium-ion batteries, each stage has been a crucial link in a chain, enabling the next big leap. Recognizing these as transitional technologies underscores the importance of continuous innovation and adaptation, with each step paving the way for the next.

This journey of progress is dynamic and complex, unfolding not in a straight line but through a web of interwoven advancements that drive society forward. EV technology exemplifies this winding path—ushering in a more sustainable future by reducing reliance on fossil fuels. Yet, as we celebrate these achievements, we must also acknowledge the limitations of current technology, which reminds us that innovation is still very much a work in progress.

The progress we see in EVs is thanks to the persistent efforts of scientists, engineers, and corporate pioneers who push the boundaries of possibility. Their work has brought us closer to a low-carbon future, reducing emissions and lessening our environmental impact. However, this success story is also a call to action: Even as we benefit from today's advancements, we must continue to seek out and build better solutions. Each breakthrough, while valuable, is ultimately just another stepping stone, driving us to innovate further and strive for a cleaner, more efficient world.

In this way, technology's journey is as much about the road ahead as it is about the road traveled—both inspiring and challenging us to keep reaching for the next great leap.

And now, artificial intelligence is that leap. Enter Google DeepMind—a force reshaping our future by accelerating material science and transforming our approach to technology's most pressing challenges.

ENTER ARTIFICIAL INTELLIGENCE:
GOOGLE DEEPMIND'S ROLE

Founded in 2010 by Demis Hassabis, Shane Legg, and Mustafa Suleyman, DeepMind began as a bold idea, merging expertise in neuroscience, computer science, and AI to solve complex global problems. When Google acquired DeepMind in 2014, this small British startup gained access to vast computational resources, giving it the fuel to pursue its ambitious goals at a larger scale.

Today, Google DeepMind is a leader in applying advanced AI to revolutionize fields critical to our future, including the development of new materials for sustainable energy solutions.

DeepMind's AI platform brings a new edge to material science by using algorithms and machine learning to analyze and predict properties of materials in record time. Traditionally, discovering new materials involves years of tedious trial and error in labs, with researchers testing endless combinations and analyzing results. DeepMind's AI, however, can screen extensive databases of chemical compounds and materials, pinpointing potential candidates with specific characteristics—such as energy density, stability, and environmental impact—at a fraction of the time.[82]

One of DeepMind's most impressive tools is GNoME (Graphical Networks for Material Exploration). This AI model has predicted the structures of 2.2 million new materials, with more than 700 of these materials synthesized and currently undergoing testing. GNoME works by combining two deep-learning models: One generates new structures from existing materials, while the other assesses stability based solely on

82 Jake Hertz, "Google, Carnegie Mellon Use AI for Battery Breakthroughs," EEPower, January 22, 2024, https://eepower.com/tech-insights/google-carnegie-mellon-use-ai-for-battery-breakthroughs/.

chemical formulas. With this process, GNoME has identified 528 promising lithium-ion conductors, offering a potential breakthrough for creating more efficient batteries.[83]

Beyond predicting new materials, DeepMind's AI is instrumental in optimizing the ones we already have, enhancing the safety and performance of existing technologies. In lithium-ion batteries, for example, DeepMind is identifying additives and alternative materials that can improve battery longevity and efficiency, paving the way for safer, longer-lasting power sources.

The GNoME project embodies the AI-driven shift in material science, aiming to lower the cost and accelerate the timeline for new discoveries. So far, 736 materials predicted by GNoME have been synthesized in the lab, demonstrating the model's accuracy and the promising future of AI-powered material discovery. This work is not just about speeding up research; it opens doors to sustainable advancements like more efficient electric vehicle batteries and superconductors for advanced computing.

AI platforms like Google DeepMind represent the next pivotal step in our technological evolution. Just as EV technology revolutionized transportation, DeepMind's work in material science could reshape energy storage and consumption. By leveraging AI, we can not only meet rising energy demands but also address sustainability challenges more effectively. Each breakthrough lays a foundation for future advancements, making AI an essential driver of innovation—and Google DeepMind, a crucial architect of tomorrow's solutions.

---

83 Amil Merchant and Ekin Dogus Cubuk, "Millions of New Materials Discovered with Deep Learning," Google DeepMind, November 29, 2023, https://deepmind.google/discover/blog/millions-of-new-materials-discovered-with-deep-learning/; June Kim, "Google DeepMind's New AI Tool Helped Create More Than 700 New Materials," *MIT Technology Review*, November 29, 2023, https://www.technologyreview.com/2023/11/29/1084061/deepmind-ai-tool-for-new-materials-discovery/.

CLOSING THOUGHTS: OUR SHARED RESPONSIBILITY

As we've seen, our rechargeable reality comes at a cost, one that we all share. From smartphones and laptops to electric vehicles, each of us contributes to the demand driving lithium and cobalt extraction—processes that often come with steep environmental and human consequences. Just as I have tested my own limits in endurance sports, pushing the boundaries of my own "batteries," we're collectively pushing the planet's resources, testing how far we can go without truly understanding the impact.

By acknowledging our impact and engaging in these vital discussions, we can influence a future where technological progress aligns with ethical and sustainable practices. It's time for us to reflect on the choices we make—choices that hold the power to benefit both humanity and the planet. Together, we can navigate the complex landscape of progress with integrity, fostering innovations that honor our responsibility to each other and to the earth.

CHAPTER 7

THE ECO PARADOX

"Tech Utopia: Just a little more science and technology and we will finally be there."

—MATT GOODMAN, PHD

As the California sunset painted the sky over the Long Valley Caldera—a dormant volcano nestled against the sharp ridges of the Sierra Nevada—I found myself enveloped in the evening alpenglow. At 7,000 feet on the north slopes of Red Mountain, just seventeen miles southeast of Mammoth Lakes, California, the tranquil scene unfolded before me. Leaning against a horse fence, I watched our gentle giant Belgian horses graze quietly, their calm presence adding to the timeless peace of the landscape.

Wrapped in the golden light of dusk, I watched as the valley below sank into shadow, slowly surrendering to the night under the towering peaks. The warm glow around me contrasted beautifully with the deepening darkness below, creating a stillness that felt like a perfect balance between light and dark—a moment so clear and sharp it matched the crispness of the mountain air.

This dance of light and shadow pulled my thoughts to the valley, where the once-steady glow from neighborhood homes now flickered with unsettling regularity. Each blackout slicing through the darkness was a stark reminder of California's strained electric infrastructure—overwhelmed by rising demand and constrained by an aging grid. The calm around me, set against this backdrop of instability, highlighted a fragile balance: the timeless serenity of nature contrasted by the creeping strain of our technology.

In a remote place like this, the electrical grid's reliability has always been a worry. But lately, with outages becoming more frequent, that worry has turned into a real and immediate concern, bringing an urgency we hadn't felt before. What used to be a "what if" now felt like a constant threat, a reminder that our grid isn't built for the demands of today. One winter, this reality hit my family and me especially hard as we watched the lights in the valley blink out one by one, leaving whole homes in darkness. It was a stark, sobering sign of a grid struggling—and often failing—to keep up.

Staring down this instability, my family made a choice to step out of the cycle. We installed a solar system with a battery backup, cutting our dependence on the grid and turning our home into its own power source. It wasn't just a practical decision; it was a statement of self-reliance in a world that feels increasingly unpredictable, turning our home into a self-reliant haven against the backdrop of a flickering valley.

This decision, though it felt like a win for independence, also made me think about the broader picture—the looming questions around California's energy future. While California—the world's fifth-largest economy—leads the charge to cut greenhouse gases and electrify everything from cars to buses, that same grid we escaped is carrying a heavier load than ever.

It's a bold goal, no question, but it forces us to ask: Can our infrastructure handle the demands of a green future?

From my perspective on my ranch, the shift toward off-grid solutions like ours makes sense for those who can make it happen. But it also points to a much bigger need for rethinking how we generate, store, and move energy across the state. If California is to meet its own ambitious goals—an 85 percent reduction in greenhouse gases and a nearly complete phase-out of fossil fuels in the next two decades—then we're looking at a transformation on an unprecedented scale.[84] A huge part of this plan is the requirement that by 2035, every new car sold must be electric.[85] That's an enormous leap in the right direction for the environment, but it also adds an equally enormous demand on an already stretched grid.

These are commendable targets, no doubt. But getting there won't be easy. The state's grid is still heavily tied to fossil fuels, and while renewable sources like wind and solar are expanding, they come with their own set of challenges. As Edward Ring from the California Policy Center points out, we face serious barriers when it comes to storing and transmitting this energy effectively. If we don't find a way to store renewable energy when the wind isn't blowing or the sun isn't shining, and if we can't improve our transmission lines to handle the load, we could be looking at even more blackouts—and on a wider scale.[86]

---

84 "California Releases World's First Plan to Achieve Net Zero Carbon Pollution," State of California Office of the Governor, accessed March 20, 2025, https://www.gov.ca.gov/2022/11/16/california-releases-worlds-first-plan-to-achieve-net-zero-carbon-pollution/.

85 Gavin Newsom, *Executive Order N-79-20* (State of California Executive Department, 2020), https://www.gov.ca.gov/wp-content/uploads/2020/09/9.23.20-EO-N-79-20-Climate.pdf.

86 Thomas Catenacci et al., "Energy Expert Sounds Alarm Over California Power Grid: 'Stressed to the Limit,'" Fox Business, February 14, 2023, https://www.foxbusiness.com/energy/energy-expert-sounds-alarm-california-power-grid-stressed-limit.

Ring's insights highlight a critical piece of California's green transition: the pressing need for robust energy storage and better distribution networks. Right now, we lack the infrastructure to store and deliver renewable energy reliably across the state. And with the demand only set to rise as electric vehicles and electrified homes become the norm, this gap is more urgent than ever.

As California charges forward on its green energy transition, a new and unexpected challenge has come knocking—a challenge that likely caught both the public and policymakers off guard. Hello, artificial intelligence, the new kid in town with a serious hunger for power. Just as we're ramping up sustainable energy sources, AI's rapid adoption and energy demands could tip our already fragile grid over the edge.

Will AI's massive appetite for electricity overwhelm our infrastructure, or could advancements in AI's own efficiency hold the key to solving our deepening energy crisis?

California, celebrated for its pioneering spirit in both tech and environmentalism, now finds itself once again in the hot seat. The state is at a critical juncture, trying to strike a delicate balance between ambitious environmental goals and the nuts-and-bolts realities of updating its outdated energy infrastructure. Leading the way is no easy task; it brings intense scrutiny, introduces endless pitfalls, and increases the risk of very public missteps. But as a trailblazer, California faces the monumental task of charting a new course into an uncharted energy future, a mission loaded with high expectations and even higher stakes.

Our state's bold drive toward a renewable grid and electrified infrastructure is unfolding against the backdrop of some of the highest electricity rates in the country. Californians paid an average of 26.52 cents per kilowatt-hour for residential electricity in 2023, a whopping 79.8 percent above the national average.[87] Why? Because maintaining and upgrading this sprawling, aging power system across our wildly diverse landscape—from the bustling streets of Los Angeles, San Diego, and the Bay Area to remote, high-desert terrain and isolated mountain homes like mine in the Sierra Nevada—is a logistical and financial nightmare.

These challenges grow even more daunting as California locks in its place as an AI powerhouse. The state is home to over 80 percent of the companies on Forbes's AI startup top 50 list, and tech hubs like the Bay Area and newer hot spots like San Diego are brimming with AI activity. San Diego, recognized by the Brookings Institution as an "early adopter" of AI, has a growing ecosystem fueled by cutting-edge research, federal contracts, and ambitious startups.[88] California's AI innovators are not only leading the charge—they're early adopters and high-volume users. AI platforms like ChatGPT, Google's Gemini, ChatSonic, and Bing AI aren't just cool new tools; they're energy guzzlers, estimated to consume 100 to 1,000 times more power than everyday digital activities like email.[89]

And more Californians are using them daily.

87 California Center for Jobs and the Economy, *California Energy Price Data for April 2023* (California Center for Jobs and the Economy, 2023), 4, https://centerforjobs.org/wp-content/uploads/April-2023-Energy-Report-.pdf.

88 Kate Murphy, "California Goes All In on AI," *Axios*, August 2, 2023, https://www.axios.com/local/san-diego/2023/08/02/california-san-diego-ai-technology-forbes-brookings.

89 Lauren Leffer, "The AI Boom Could Use a Shocking Amount of Electricity," *Scientific American*, October 13, 2023, https://www.scientificamerican.com/article/the-ai-boom-could-use-a-shocking-amount-of-electricity/; Delger Erdenesanaa, "A.I. Could Soon Need as Much Electricity as an Entire Country," *New York Times*, October 10, 2023, https://www.nytimes.com/2023/10/10/climate/ai-could-soon-need-as-much-electricity-as-an-entire-country.html; Sarah Wells, "Generative AI's Energy Problem Today Is Foundational," IEEE Spectrum, October 29, 2023, https://spectrum.ieee.org/ai-energy-consumption; Nathi Magubane, "The Hidden Costs of AI: Impending Energy and Resource Strain," Penn Today, March 8, 2023, https://penntoday.upenn.edu/news/hidden-costs-ai-impending-energy-and-resource-strai; Reece Rogers, "AI's Energy Demands Are Out of Control. Welcome to the Internet's Hyper-Consumption Era," *Wired*, July 11, 2024, https://www.wired.com/story/ai-energy-demands-water-impact-internet-hyper-consumption-era/.

This explosive AI growth in California comes right as we're pushing aggressively to ditch fossil fuels, putting even more pressure on a grid already struggling to meet demand. AI's rising energy consumption piles an extra load onto a system in flux—a transition weighed down by logistical and financial challenges. With San Diego and the Bay Area leading the charge in AI, California's energy infrastructure must not only keep up with demand but adapt to support these power-hungry advances.

Navigating this complex terrain will require a nuanced approach to juggle California's commitments to both tech innovation and environmental sustainability. The state's pioneering efforts in renewable energy and AI can't exist in silos; they're deeply interconnected. If California is to successfully carve out a sustainable energy future, it will need innovation and infrastructure to work hand in hand, influencing the affordability and reliability of power across the state.

In the grand scheme of things, California's dual role as tech pioneer and electrification trailblazer is veering dangerously close to a head-on collision with its grid's capacity and the financial reality facing both taxpayers and the state's budget. How California navigates this balancing act will shape whether its ambitious energy goals can coexist with a dependable power supply.

The whole world is watching.

And yet, within this pressure-cooker environment, AI—despite its massive energy needs—might also be the very tool we need to increase energy efficiency. How's that for a paradox? The technology that strains our energy resources could be the one that ultimately helps us manage them better. For Californians, this creates a fascinating "eco paradox": the very AI advances adding pressure to the grid might also deliver solutions to help stabilize it.

AI isn't just a formidable obstacle; it has the potential to drive a major shift in energy sustainability in California. This perspective places AI squarely at the heart of both the technological solutions and the ethical questions that will define our future.

In a broader context, California's leadership in tech and commitment to electrification doesn't just highlight local challenges—it reflects a global dilemma. Around the world, the race to innovate and reduce carbon footprints collides with the hard limits of infrastructure and economics. It's the ultimate paradox: Technologies that stress our energy resources might also be the ones to help save them. Investigating how AI can coexist—and even actively contribute—to a sustainable future presents a thrilling possibility, casting AI not merely as an energy-hungry challenge but as a catalyst for reshaping energy use and sustainability on a global scale, with California leading the charge.

Can AI truly be a solution to environmental challenges when it's also part of the problem? In this chapter, we'll explore the "eco paradox" of artificial intelligence—its dual role as both a powerful tool for environmental sustainability and a substantial energy consumer with a hefty carbon footprint. Through a series of real-world examples, you'll learn how AI is being harnessed to drive efficiency in sectors like agriculture, water resource management, and renewable energy, showcasing its transformative potential. At the same time, we'll address the thorny side of AI: the resource strain and environmental impact that come with the rapid expansion of this technology. By examining both the roses and thorns of AI's development, this chapter aims to reveal how responsible innovation, transparency, and digital sobriety can help strike a balance between AI's remarkable capabilities and the sustainability challenges it presents.

> **The Paradox:** AI increases energy efficiency but also drives massive energy consumption, straining the grid and amplifying carbon emissions.
>
> **The Roses:** AI optimizes energy use and enables renewable energy integration.
>
> **The Thorns:** Training and running AI models consume enormous amounts of electricity, offsetting some sustainability gains.

THE THORNS
AI'S INSATIABLE ENERGY DEMAND

The explosive growth in AI is more than just a tech revolution—it's an energy-intensive transformation with deep environmental implications. Our daily routines, from streaming movies and chatting with digital assistants to running smart home devices, all depend on powerful AI-driven systems. But this convenience comes at a cost: The vast backend of data centers and network infrastructure that support these technologies guzzle enormous amounts of energy. As our dependence on digital tools grows, so does our responsibility to consider how these habits drive demand—and to push for more energy-efficient AI and sustainable tech practices.

According to the Shift Project, a French nonprofit advocating for a post-carbon world, the gains in energy efficiency achieved every 1.52 years through Koomey's Law (which describes computing efficiency improvements) are being eclipsed by the exponential rise in data consumption.[90] From

---

90 "The Shift Project." https://theshiftproject.org/en/home/; Jonathan Koomey et al., "Implications of Historical Trends in the Electrical Efficiency of Computing," *IEEE Annals of the History of Computing* 33, no. 3 (March 2011): 46–54, https://doi.org/10.1109/MAHC.2010.28.

streaming services to cloud gaming, our digital appetites are expanding faster than efficiency improvements can compensate, pushing up both energy consumption and emissions. This growth isn't just a numbers game; it's a physical expansion that includes more data centers, devices, and complex networks—all leaving a substantial environmental footprint. Streaming video, which uses AI to optimize quality and traffic flow, is one of the most variable factors in energy consumption. Everything from the data center's efficiency to user behavior impacts the power required for streaming. Globally, data centers now consume around 1 percent of all electricity. But as internet traffic has tripled and data center workloads have more than doubled since 2015, this figure is primed to rise, despite shifts toward more efficient and hyperscale data centers. Currently, digital technologies account for nearly 4 percent of global emissions, and the numbers are climbing fast. Streaming media, for example, is ballooning: In 2019, Netflix subscriptions grew by 20 percent, and the electricity consumption linked to streaming surged by 84 percent. New AI-driven services, especially in video streaming and cloud gaming, are appearing at an unprecedented pace. Mobile video traffic in particular is booming, growing at 55 percent annually. Today, mobile devices account for over 70 percent of the billion hours of YouTube content streamed every day.[91]

These trends have led the Shift Project to caution that if the current growth in digital demand and AI infrastructure continues unchecked, we'll see significant environmental

91   Maxime Efoui Hess and Jean-Noël Geist, *Did the Shift Project Really Overestimate the Carbon Footprint of Online Video?: Our Analysis of the IEA and CarbonBrief Articles* (The Shift Project, June 2020), 25–27, https://theshiftproject.org/wp-content/uploads/2020/06/2020-06_Did-TSP-overestimate-the-carbon-footprint-of-online-video_EN.pdf; George Kamiya, "Factcheck: What Is the Carbon Footprint of Streaming Video on Netflix?," CarbonBrief, February 25, 2020, https://www.carbonbrief.org/factcheck-what-is-the-carbon-footprint-of-streaming-video-on-netflix/.

repercussions. They call for a strategy of "digital sobriety," a deliberate balancing act between tech innovation and responsible energy use. The goal is clear: manage the digital sector's environmental impact by aligning technological advances with strong climate policies that can keep pace with global energy and emissions targets. Such an approach is crucial if we want digital services to serve society without leaving a lasting mark on the planet.

Data centers, critical to supporting AI, have seen a 25 percent annual growth rate in electricity consumption from 2005 to 2021, far outpacing the 7 percent growth in renewable energy investment. Generative AI models like ChatGPT amplify this energy appetite. Launched by OpenAI on November 30, 2022, ChatGPT achieved a jaw-dropping adoption rate, surpassing a million users within just five days. Fast-forward to March 2025, and the OpenAI website hosted 5.19 billion visits in a single month.[92] AI-driven platforms like ChatGPT require vast computational resources, with training a single large language model emitting as much carbon as five cars over their entire lifetimes. This immense energy demand highlights the environmental cost of developing and running these technologies at scale.

Additionally, ChatGPT's reach on social media, particularly on YouTube, speaks volumes about the cultural draw of AI. Around 65 percent of ChatGPT's social media traffic comes from YouTube, reflecting a strong interest in AI-driven content. Approximately 12 percent of its users are based in the United States, further illustrating its widespread appeal. With about 100 million weekly users, ChatGPT has quickly become indis-

---

92   Fabio Duarte, "Number of ChatGPT Users (March 2025)," *Exploding Topics* (blog), last updated March 25, 2025, https://explodingtopics.com/blog/chatgpt-users.

pensable for everything from casual queries to professional applications.

THE ROSES

AI AND SUSTAINABILITY: THE IRONY OF INNOVATION

While AI's skyrocketing energy demands pose a clear challenge to environmental sustainability, this very technology holds immense promise for advancing efficient energy use and supporting green initiatives.

The irony isn't lost on us: AI requires colossal computing power, often supplied by carbon-intensive energy sources, which increases the carbon footprint of tech. Yet, as AI continues to bloom, it has the potential to drive smarter energy management, from streamlining data center power use to improving fuel efficiency in transportation. This dual nature of AI presents a complex but compelling opportunity—AI is not only part of the sustainability problem but could also be the key to solving it.

The good news: The tech world is already taking action. In 2022, major players like Amazon, Meta, and Google contracted a whopping 77 gigawatts of renewable energy to fuel their operations—a capacity sufficient to power thousands of data centers.[93] This commitment to sustainable energy sources like solar, wind, and hydro marks a significant shift in the backbone of data center infrastructure. Google and Microsoft are pushing even further, pledging to match their total energy con-

---

93   Arghanshu Bose, "Tech Giants and Clean Energy: Amazon, Meta, Google Biggest Buyers Among Other Companies," *Times of India*, January 19, 2023, https://timesofindia.indiatimes.com/gadgets-news/tech-giants-and-clean-energy-amazon-meta-google-biggest-buyers-among-other-companies/articleshow/97122006.cms; Catherine Clifford, "Clean Energy Amazon, Meta and Google Buy More Clean Energy Than Any Other Companies," CNBC, January 18, 2023, https://www.cnbc.com/2023/01/18/amazon-meta-and-google-buy-more-clean-energy-than-any-other-companies.html.

sumption with carbon-free sources by 2030. This bold move not only targets the tech sector's environmental footprint but also opens doors for job creation within the clean energy industry.[94]

On top of shifting energy sources, tech companies are exploring more energy-efficient hardware for AI computations and even looking into quantum computing for its potential to revolutionize energy use. AI is also being employed to optimize energy use within the data centers themselves and across smart grids, maximizing efficiency from within.

To get a sense of AI's energy footprint, let's consider Benjamin Lee's study on large language models (LLMs), which shows that energy use is spread across three main phases: preprocessing (30 percent), training (30 percent), and serving (40 percent). As models evolve, improvements in the preprocessing and training phases could bring down their energy demands, marking a step toward a more sustainable AI.[95]

And then, there's the role of clean energy technologies. Despite its controversies, nuclear power is back on the table for its low greenhouse gas emissions, with newer reactor designs that promise greater safety and efficiency.[96] Looking even further ahead, fusion energy—still in the early stages of development—could one day provide a massive, clean power source for high-demand technologies like AI. With AI aiding in the advancement and safe deployment of these technologies,

94 Andrew Blum, "How Amazon Became the Largest Buyer of Renewable Energy in the World," *Time*, September 15, 2022, https://time.com/6213666/amazon-renewable-energy/; "Net-Zero Carbon," Google Sustinability, accessed March 20, 2025, https://sustainability.google/operating-sustainably/net-zero-carbon/; Brad Smith, "Microsoft Will Be Carbon Negative by 2030," *Official Microsoft Blog*, January 16, 2020, https://blogs.microsoft.com/blog/2020/01/16/microsoft-will-be-carbon-negative-by-2030/.

95 Ferre, I. "Energy Consumption 'to Dramatically Increase' Because of AI." *Yahoo Finance*, September 30, 2023.

96 Drake Bennett, "Microsoft Sees Artificial Intelligence and Nuclear Energy as Dynamic Duo," *Bloomberg*, September 29, 2023, https://www.bloomberg.com/news/newsletters/2023-09-29/microsoft-msft-sees-artificial-intelligence-and-nuclear-energy-as-dynamic-duo.

the energy landscape for AI could be transformed in ways that align with a sustainable future.

AI is reshaping the way industries approach resource management, opening new doors for environmental sustainability. With the power to analyze vast datasets, streamline operations, and generate innovative solutions, AI is making a tangible impact on reducing environmental footprints across essential sectors like energy, agriculture, water management, and renewable energy.

In the following examples, we'll explore how AI is being deployed—from Google DeepMind's groundbreaking work in data center cooling to IBM Watson's advancements in smart agriculture, Aquaai's robotic fish for real-time water monitoring, and Google's Project Sunroof, which helps maximize solar power potential. These applications showcase AI's remarkable potential not only to reduce resource consumption but also to drive sustainable progress. And as AI technology advances, its role in building a more sustainable future will only grow.

AI Optimizing Resource Usage

In the last chapter, we explored how Google DeepMind is breaking new ground in material science by helping to source alternatives to lithium and cobalt—materials critical to building a more sustainable rechargeable battery. This example is just a glimpse of how AI is fast becoming a cornerstone in the drive toward environmental sustainability. With its unmatched ability to optimize resource use, enhance data analysis, and inspire innovations, AI is transforming efforts to reduce our environmental impact.

One of the most compelling areas where AI is making strides is in the energy sector, especially in large-scale operations like data centers. Take Google's DeepMind project again. When Google needed to tackle the massive energy costs associated with cooling its data centers, DeepMind's AI was deployed to find a solution. Data centers, which host the servers and infrastructure that keep internet and cloud services running, burn through vast amounts of energy just to stay cool. The sheer heat generated by the servers requires constant cooling, making it one of the most energy-intensive aspects of data center operations.

DeepMind's approach was smart and data-driven. The AI system analyzed streams of data from thousands of sensors throughout the data center—tracking variables like temperature, pressure, and humidity. Using machine learning, DeepMind detected patterns and predicted shifts in these conditions, allowing the AI to suggest optimal cooling strategies that adjust to real-time needs. The result? DeepMind's AI cut the energy used for cooling by an impressive 40 percent, all while ensuring that data center operations ran smoothly.[97]

This substantial reduction showcases not only how AI can boost energy efficiency but also how it can lessen the environmental impact of tech infrastructure. By rethinking how resources are used, AI is not only saving energy but also setting an example for how other industries can harness intelligent systems to reach sustainability goals. As AI technology advances and integrates further into various sectors, its role in supporting a greener future is expected to grow, unlocking new ways to soften the environmental impact of our daily lives.

97   Richard Evans and Jim Gao, "DeepMind AI Reduces Google Data Centre Cooling Bill by 40 Percent," *DeepMind* (blog), July 20, 2016, https://deepmind.google/discover/blog/deepmind-ai-reduces-google-data-centre-cooling-bill-by-40/.

AI and Sustainable Agriculture

As the world's population grows and climate change intensifies, sustainable farming practices are more critical than ever. IBM's Watson Decision Platform is at the forefront of this movement,[21] using AI and other advanced technologies to help farmers optimize resource use, boost productivity, and reduce environmental impact.

IBM's Watson Decision Platform is transforming agriculture by harnessing the power of AI, weather data, the Internet of Things (IoT), and blockchain technology to provide farmers with real-time, actionable insights.[98] This cutting-edge platform gathers and analyzes data from multiple sources—including weather forecasts, soil moisture, and crop performance—tailoring insights to meet the unique needs of each farm. For example, using predictive analytics, it offers highly accurate weather forecasts, allowing farmers to plan planting and irrigation schedules to avoid adverse weather impacts. Soil health is continuously monitored through sensors that analyze moisture and nutrient levels, helping farmers make informed decisions about fertilization and irrigation and reducing the risk of resource overuse.

The platform also uses AI-driven image recognition and sensor data to detect early signs of crop stress or pest infestations. By identifying these issues early, farmers can intervene promptly, reducing the need for pesticides and minimizing crop damage.

An added layer of innovation comes from blockchain technology, which enhances transparency and traceability across the agricultural process. This secure, immutable ledger

---

98   Dan Wolfson, *IBM Watson Decision Platform for Agriculture: Using A.I. to Aid in Decision Making from Farm to Fork* (IBM, 2019), 2, https://worldagritechusa.com/wp-content/uploads/2019/03/Dan-Wolfson-IBM.pdf.

records all data inputs and AI recommendations in a chain of blocks, making it nearly impossible to alter the data without consensus from the network. Blockchain creates a reliable audit trail, allowing for clear accountability from farm to table and reinforcing consumer trust in the supply chain. By integrating blockchain, IBM's platform not only supports smarter farm management and regulatory compliance but also builds transparency, offering a full record of every action—from soil treatment to harvesting.

As AI technology and its tools advance, agriculture may see even more innovative applications, such as autonomous machinery and robotic harvesting, paving the way for enhanced productivity and sustainability in farming.

AI and Water Resource Management

Aquaai, an innovative startup, is tackling water resource management with a unique approach: AI-powered robotic fish. These robotic fish are designed to blend seamlessly into aquatic environments, gathering critical data without disrupting natural habitats or interfering with commercial activities. Outfitted with advanced sensors, they navigate diverse water bodies—from commercial fish farms and natural lakes to rivers and industrial sites—monitoring key parameters like pollution levels, water temperature, and salinity.

The insights gathered by Aquaai's robotic fish are invaluable for managing water resources more effectively. In commercial fish farms, for example, the data helps maintain ideal water conditions, supporting healthy fish growth and reducing disease risks. In natural bodies of water, these robots track environmental conditions over time, detecting pollutants or other threats to aquatic life and alerting authorities to act before problems escalate. This capability plays a vital role in preventing overfishing and protecting habitats from environmental degradation.

In industrial settings, these robotic fish provide real-time monitoring of waste discharge and its impact on water quality, ensuring compliance with environmental regulations and helping to minimize industrial impacts on natural water reserves. By leveraging AI to analyze this wealth of data, Aquaai's solution not only improves immediate water management but also supports long-term conservation efforts with a continuous stream of accurate, actionable insights.

This approach exemplifies how AI can revolutionize environmental management, turning tasks that once required extensive human effort into efficient, automated processes—a promising step toward more sustainable water resource management worldwide.

AI and Solar Power Optimization

As cities strive to reduce their carbon footprints, solar energy is emerging as a key solution—and Google's Project Sunroof is making it easier than ever for property owners to join the movement.[99]

Google's Project Sunroof is an innovative AI-driven tool that uses satellite imagery to evaluate the solar potential of rooftops in urban areas. This sophisticated technology helps homeowners and businesses quickly assess whether their buildings are suitable for solar panel installation. By simply entering an address into Project Sunroof, users receive a detailed, AI-powered analysis of their roof's solar capacity, including

---

99   Ana Almerini, "What Is Google Project Sunroof?," *SolarReviews* (blog), accessed March 20, 2025, https://www.solarreviews.com/blog/what-is-google-project-sunroof.

sunlight exposure throughout the year and potential energy output from solar panels.

The AI behind Project Sunroof considers various factors—roof orientation, shade from nearby structures and trees, and local weather patterns—to provide accurate and personalized recommendations for optimal solar panel placement. This makes it easier for property owners to make informed decisions about solar energy, knowing exactly how much energy their setup could generate.

Beyond simplifying the decision-making process, Project Sunroof also optimizes solar panel configurations to maximize energy output. The tool even estimates potential cost savings, giving urban residents and businesses an added incentive to invest in renewable energy. By offering easy access to this critical information, Google's Project Sunroof is helping to accelerate the shift to solar power in cities, supporting energy sustainability and reducing greenhouse gas emissions. This approach highlights how AI can drive environmental sustainability and empower people and businesses to make smart, eco-friendly choices that align with broader climate goals.

CLOSING THOUGHTS: THE DANCE OF ROSES AND THORNS

As we've seen, the relationship between AI and environmental sustainability is a blend of roses and thorns—a delicate balance between opportunity and responsibility. On one hand, AI holds the potential to transform our approach to energy use, resource management, and conservation, much like a keystone species that supports an entire ecosystem. Its innovations can power efficient solar grids, monitor water quality, and even revolutionize agriculture, becoming essential tools in our climate resilience efforts.

Yet, these advances come with significant challenges: the thorny reality of AI's energy demands, carbon footprint, and ethical implications, which can disrupt the very sustainability we seek. Striking this balance requires transparency, responsible innovation, and vigilant management of AI's life cycle from production to disposal. As AI becomes more embedded in daily life, the call for "digital sobriety" becomes even more urgent, suggesting that the solutions to our digital age may also lie in rethinking how we consume and build these technologies. By making thoughtful, sustainable choices, we can help ensure that our technological advances serve society without overwhelming the planet.

In the end, just as ecosystems flourish with careful stewardship, our use of AI must be guided by a commitment to harmony with the environment. By nurturing this balance, we can create a future where AI and the natural world coexist in a symbiotic dance—where the roses of intelligent innovation flourish alongside the thorns of its challenges, enriching our planet without undermining it.

CHAPTER 8

AI AND THE FUTURE OF WORK

"What is hard for humans is easy for machines, and what is easy for humans is hard for machines."

—Moravec's Paradox, Hans Moravec,
Carnegie Mellon University, 1988

"Humans won't be replaced by AI. Humans will be replaced by humans using AI."

—Kyle Forest, US Human Capital Chief Marketing
Officer. Deloitte EMTech Digital Conference. MIT
Campus. Cambridge, Massachusetts, May 12, 2024

As dawn broke over the fields, an age-old scene of enduring strength and tradition came alive.

The air carried the earthy scent of freshly cut hay mixed with the sweet smell of Belgian draft horses. For those who may not know, Belgian horses are different from other horses. They smell different. They feel different.

More than anything, these gentle giants are bred for hard work.

Their massive frames and kind eyes expressed patience as they flicked away summer flies. Their soft lips quivered in the cool morning air, eager for the tasks ahead. For them, cutting, raking, and baling hay weren't just chores—they were a vital part of their being.

Hay harvesting has been a part of human life for over 10,000 years, originating in ancient regions like Turkey, Syria, and Iraq, where it was key to the domestication of livestock. It's a grueling task, whether performed today or millennia ago. From the outside, it might seem like an overwhelming test of endurance, but the experience is far from that. In the fields, modern distractions like cell phones fall away, leaving only the shared goal of the family, purpose, and the hard work that binds them. The truth is, laboring together doesn't break us—it strengthens us, as much today as it did ten thousand years ago.

The introduction of draft horses transformed hay harvesting, representing a leap in productivity. First domesticated thousands of years ago, these powerful animals were widely used in Europe by the medieval period. The synergy between horse strength and human intellect revolutionized agriculture, easing the burden on people and boosting efficiency. This partnership—man and horse working together—symbolized a harmonious blend of nature and human endeavor, creating a bond that has shaped agricultural practices for centuries.

The adoption of draft horses wasn't just about productivity; it spurred advancements in other fields, too. Breeding stronger horses, creating durable harnesses, forging horseshoes, and building barns all required new skills and knowledge. This integration of horses into daily work didn't just multiply output—it expanded the overall knowledge base of society. Horse farming became a transitional technology that inspired modern, machine-driven farming.

In 2015, my wife and I moved our family from California to Walker's Creek, Virginia, where we bought a farm on two hundred acres. Instead of depending on machines, we chose to do things differently. We harvested hay and logged our forests with horses, not tractors. Our goal was to develop family values of hard work, foster the skillset and ability to fix things, and build teamwork toward a common goal—an education for the entire family. My wife and I decided to do this for several reasons: to teach our kids about history (and actually live it), introduce challenges, and finally—and most importantly—do it all for fun.

Things were going well up until 2018, when I was hit by a distracted driver while cycling. This incident left me unable to work with the horses or handle the farm's daily tasks. This forced us to adopt tractors and other mechanized equipment made to make life easier, for both the horses and me. At first, it appeared that the horses appreciated the break from their strenuous duties. Certainly I did as I healed from my injuries. However, as time passed, the absence of regular work took its toll on them, and possibly on me as well. The once vivid spark in their eyes dimmed, their robust frames began to soften, and their demeanor shifted, becoming characterized by random and unpredictable bouts of lethargy and irritability.

This change was a poignant reminder of how deeply their sense of purpose was tied to their daily task, which then got me thinking: What will happen to us when work—something that has defined us for generations—is no longer required? Are we humans any different from horses? As much as we daydream about easier lives, doesn't our work give us purpose?

Will we, like my horses, grow fat, lethargic, and irritable when our work is gone?

In this chapter, we'll explore how AI technology offers convenience and efficiency (the roses) while at the same time, how it also threatens our sense of purpose (the thorns). Just as my horses lost their vitality without their work, humans may face existential crises as AI reshapes our roles, too. Are we ready for the changes AI will bring? And how can we address the potential loss of purpose?

The Paradox: AI can enhance human productivity but also risks displacing workers and devaluing certain skills.

The Roses: Automation creates opportunities for efficiency and innovation.

The Thorns: Widespread job disruption and inequality.

Before diving in, a quick note on structure: Unlike the rest of the chapters in Part Two, where the roses and thorns of AI's impact are clearly separated, this chapter weaves them together. That's because the very nature of work—and the role AI plays in it—is deeply intertwined. The opportunities AI presents and the risks it poses are not easily divided; instead, they unfold together, shaping our evolving relationship with labor, purpose, and technological progress.

MEET BELLABOT

In 2024, we gathered at a bustling sushi restaurant in Lake Havasu, Arizona, to celebrate my youngest daughter Liv's sixteenth birthday. Our table buzzed with the excitement of her newly earned driver's license and the sense of growing adulthood for her and her friends. My wife, Kim; our other children,

Mia (twenty-one) and Dane (eighteen); and our family friend McKenna (sixteen) all shared in the joy of the evening. Our waiter, Spencer, a young man clearly in his first job, embodied the enthusiasm of someone eager to please. His conversational skills and professionalism, despite him being too young to serve alcohol, impressed us all and added a personal touch to the night. Having waited tables myself to pay for school, I appreciated Spencer's effort and the warmth of human service. In my mind, I'd already decided he was getting a big tip. But amid the clinking of glasses and lively chatter, something caught my attention—a striking paradox. Gliding through the restaurant with blinking lights and effortless grace was BellaBot, a robotic server. This sleek, efficient machine was handling meals for four tables at once, operating with the precision of a seasoned waiter.

I caught Spencer's eye for a moment—was it disdain, or maybe a flicker of concern about his job?

Perhaps the bigger and more immediate question for Spencer was something different: *Who's going to get my tip? Me or Bella?*

Curiosity got the best of me, so I did a little research about BellaBot and the potential cost savings for restaurant businesses.

As of March 2025, ZipRecruiter reported that restaurant owners in Arizona typically pay around $2,188 per month in wages for a server, based on an hourly rate of $12.62 for the state.[100] (Arizona ranks on the low end, coming in at the lowest quartile out of fifty states in terms of waitstaff pay, so most states pay more.)

Contrast that with BellaBot, a robotic server costing approximately $15,900 for a one-time purchase. With a twelve-hour

---

100 "Restaurant Server Salary in Arizona," ZipRecruiter, accessed March 20, 2025, https://www.ziprecruiter.com/Salaries/Restaurant-Server-Salary--in-Arizona.

battery life, BellaBot doesn't need breaks for food, restroom visits, or coffee. There's no overtime, workers' compensation, or injury concerns. Able to carry up to eighty-eight pounds and with a lifespan of five to twenty years—about 100,000 hours of use—this robot could pay for itself in as little as six months.[101]

That's a pretty attractive value proposition for restaurant owners.

BellaBot represents the new era of service: robots efficiently managing multiple tables without the hiccups of illness, vacation days, or the need for overtime pay. Advocates argue these robots could even increase human servers' earnings by freeing them up for more customer interaction. Indeed, this technological marvel highlights many roses. Yet, this innovation also comes with a thorny dilemma: The rise of robots in service roles poses a significant threat to entry-level jobs—the very stepping stones many young adults rely on.

These jobs offer more than just a paycheck; they are critical for developing public speaking, stress management, and teamwork skills. As I watched Spencer work, I couldn't help but think about a future where opportunities like his, and those my own kids rely on, might dwindle. My twenty-one-year-old daughter, Mia, works as a barista and waitress to support herself through school and travel. Those jobs have given her confidence and independence—skills that go beyond anything we could teach her at home or in school. Entry-level jobs, it seems, are vital for society as a whole.

I thought back to our Belgian draft horses. When we transitioned from horse-drawn farming to machines, the physical and emotional toll on our horses was undeniable. Stripped of their roles, they lost their vitality and their demeanor changed.

---

101 Pudu Robotics. https://www.pudurobotics.com/.

The same risks come with automation in human jobs. This stark realization of technology's double-edged sword resonated profoundly with me as a father, prompting a critical question: As machines and AI replace the mundane tasks that once gave us purpose, how do we ensure that our youth still gain the lessons and personal growth these early job experiences provide?

The challenge is to foster an environment where technological advancements and human development can thrive together in synergy—ensuring that the advancements in AI do not eclipse the essential interpersonal skills that are nurtured through conventional work experiences.

THE ETHICAL DILEMMA OF AI-DRIVEN JOB DISPLACEMENT

The impact of AI extends far beyond entry-level workers. It reaches into vulnerable groups—older workers nearing retirement, those with less education, and people in lower-wage jobs.

AI and automation are increasingly poised to render certain jobs, especially those involving routine and manual tasks, obsolete. Entire sectors—manufacturing, service industries, and more—are at risk.[102] For many of these workers, transitioning to new roles isn't just difficult, it's nearly impossible without retraining, which they may not have the resources or opportunities to access.[103] The financial and social stability of

---

102 Behnam Tabrizi and Babak Pahlavan, "Companies That Replace People with AI Will Get Left Behind," *Harvard Business Review*, June 23, 2023, https://hbr.org/2023/06/companies-that-replace-people-with-ai-will-get-left-behind; Natalie Rose Goldberg, "'AI Exposure' Is the New Buss Term to Soften Talk About Job Losses. Here's What It Means," CNBC, October 27, 2023, https://www.cnbc.com/2023/10/27/ai-exposure-is-new-buzz-term-for-softening-talk-about-job-losses.html.

103 Jorge Tamayo et al., "Reskilling in the Age of AI," *Harvard Business Review*, September–October 2023, https://hbr.org/2023/09/reskilling-in-the-age-of-ai.

these individuals is on the line, and with it, the growing divide between social classes could widen even further.

But it's not just entry-level workers and vulnerable groups that will be displaced.

It could be me, too.

As a surgeon, I've spent decades in school and years in the operating room honing a highly specific skillset. The idea that AI could potentially displace my work used to seem far-fetched, but it's no longer a distant possibility. The rapid integration of AI into surgery is both awe-inspiring and somewhat unsettling.

After nearly three decades of performing surgeries, I've seen firsthand how robotics are enhancing surgical outcomes. These machines are capable of a level of precision that even my most experienced colleagues and I can't always match. AI-driven suture robots can autonomously close wounds, potentially replacing roles that have traditionally required skilled human assistants.

In February 2024, *MIT Technology Review* showcased an example of AI's growing presence in the operating room. Researchers at UC Berkeley demonstrated a two-armed robot that stitched six stitches on imitation skin, transferring the needle between its limbs with perfect suture tension. The robot wasn't just following instructions—it was performing the task autonomously with a surgeon's precision.[104] This marks a shift toward fully autonomous surgical robots that could soon take on intricate tasks once reserved for the most skilled human hands.

While robotic assistance in surgeries like hernia repairs or coronary bypasses has traditionally been supportive, new developments like this could revolutionize operating room dynamics.

---

104 James O'Donnell, "Watch This Robot As It Learns to Stitch Up Wounds," *MIT Technology Review*, February 22, 2024, https://www.technologyreview.com/2024/02/22/1088780/watch-this-robot-as-it-learns-to-stitch-up-wounds/.

As Ken Goldberg, the director of the Berkeley lab developing this technology, points out, suturing is one of the most complex tasks in robotics.[105] Mastering this could open the floodgates for robots performing more intricate procedures.

AI's influence isn't limited to surgery, either. Robotic IV automation (RIVA) systems are improving patient safety and cost efficiency in drug delivery, while robots like "SAM" are assessing fall risks. AI-driven responses in healthcare are also gaining favor. Patients, in a 2023 study by *JAMA*, preferred AI responses over human ones 78.6 percent of the time.[106] AI's ability to deliver detailed, empathetic, and timely feedback is appealing, especially compared to overburdened human doctors.

In a pivotal 2023 study led by Dr. Dembrower and published in *The Lancet*, AI matched radiologists in diagnostic accuracy for mammography screenings, improving both efficiency and patient experiences.[107]

The financial implications of integrating AI into healthcare are significant, especially in light of current trends in its spending. According to projections from the CMS, US healthcare expenditures reached a staggering $4.9 trillion in 2023, averaging $14,570 per person, and representing 17.6 percent of the nation's GDP.[108] With healthcare costs expected to surpass 20 percent of GDP by 2030, the cost-saving potential of AI is

---

105 O'Donnell, "Watch This Robot."

106 John W. Ayers et al., "Comparing Physician and Artificial Intelligence Chatbot Responses to Patient Questions Posted to a Public Social Media Forum," *JAMA Internal Medicine* 183, no. 6 (2023): 589–596, https://doi.org/10.1001/jamainternmed.2023.1838.

107 Karin Dembrower et al., "Artificial Intelligence for Breast Cancer Detection in Screening Mammography in Sweden: A Prospective, Population-Based, Paired-Reader, Non-Inferiority Study," *The Lancet* 5, no. 10 (October 2023): E703–E711, https://doi.org/10.1016/S2589-7500(23)00153-X.

108 "NHE Fact Sheet," Centers for Medicare and Medicaid Services, last modified December 18, 2024, https://www.cms.gov/data-research/statistics-trends-and-reports/national-health-expenditure-data/nhe-fact-sheet.

becoming crucial to maintaining the financial sustainability of the healthcare system. The potential savings AI offers in healthcare far outweigh those seen in other industries, such as the food and hospitality sector, where robots like BellaBot are already demonstrating their economic impact.

This ongoing integration of AI forces us to confront a fundamental issue: How do we balance the technological efficiency AI brings with the irreplaceable value of human purpose in work?

Once again, I find myself asking: Are we humans any different from horses?

FREEDOM OR PURPOSE LOST?

Ezra Klein explored this idea in his podcast *Will AI Break the Internet? Or Save It?*, where he discussed the nature of modern work with Nilay Patel.[109] Patel, editor in chief of *The Verge*, points out that a lot of our work—emails, reports, presentations—falls into the category of "middling" work. It's not groundbreaking, but it keeps things running. Think about it—when was the last time your PowerPoint presentation was in the running for a Nobel Peace Prize? Even the best workers spend most of their time doing this "middle C+" work, which is precisely the kind of thing AI is poised to take over.

AI optimists argue that this will free us to focus on more creative and fulfilling pursuits. But there's a flipside: Could it also eliminate jobs altogether, eroding the sense of purpose that comes from work? This is the paradox at the heart of AI. It could unleash new realms of creativity—or upend the very fabric of employment (and our purpose) as we know it.

---

109   Ezra Klein, host, *The Ezra Klein Show*, "Will A.I. Break the Internet? Or Save It?," *New York Times*, April 5, 2024, https://www.nytimes.com/2024/04/05/opinion/ezra-klein-podcast-nilay-patel.html.

In the April 13, 2024 issue of *The Economist*, an article titled "Utopian Dystopia" asked a provocative question: Would it be good for society if AI replaced all human work?[110] This conversation echoes predictions made nearly a century ago by British economist John Maynard Keynes (1883–1946), the founder of modern macroeconomics. In his essay "Economic Possibilities for Our Grandchildren," Keynes envisions a future where technological advancements and increased productivity allow people to work just fifteen hours a week—achieving the same output as someone working forty hours a week in the early twentieth century.[111]

So how accurate was Keynes's 1930 prediction?

Over the past 150 years, the average workweek has indeed shrunk significantly. Modern workers log twenty to thirty fewer hours each week compared to their nineteenth-century counterparts. This reduction is due to several factors. Technological advancements and improved efficiency have allowed us to complete the same amount of work in far less time. Changes in regulations also played a role, with laws that limit working hours to protect workers' health and quality of life. The introduction of the eight-hour workday in the early twentieth century became standard, along with the rise of holidays and paid vacations, reflecting a cultural shift toward valuing leisure time.

As national incomes grew, so did wealth distribution, which allowed for shorter working hours while maintaining or even improving living standards. Back in 1870, early industrial workers put in more than three thousand hours annually, or sixty

110 "Utopian Dystopia: What Will Humans Do If Artificial Intelligence Solves Everything?," *The Economist* (April 13–19, 2024): 62.

111 John Maynard Keynes, *Essays in Persuasion* (W. W. Norton & Co., 1963), 369.

to seventy hours per week.[112] By 1905, the eight-hour day had become widespread, and by 1926, Henry Ford set the precedent of a five-day, forty-hour workweek for his employees.[113] Fast-forward to 2023, and it was estimated that to produce the same output as a forty-hour workweek in 1950, a worker would only need to work eleven hours per week—showcasing the dramatic strides in productivity and our evolving relationship with work.[114]

Keynes, as it turns out, was amazingly accurate. Not only was he a brilliant economist, but his foresight bordered on prophetic. Interestingly, he also warned us about the human condition in a world without work. He noted that no country or people could approach an age of abundance and leisure without some unease, famously stating, "For we have been trained too long to strive and not to enjoy."[115] His words evoke a biblical notion: "Idle hands are the Devil's workshop," hinting at the fear that without work, we might lose our sense of purpose.

This historical perspective offers a valuable lens for understanding our current trajectory with AI. While the technology promises to reduce the hours we need to work, it also challenges our deeply ingrained cultural values about the nature and purpose of labor.

Simply put: Humans, much like horses, are meant to work.

---

112 Charlie Giattino and Esteban Ortiz-Ospina, "Are We Working More Than Ever?," Our World in Data, December 16, 2020, https://ourworldindata.org/working-more-than-ever.

113 Erica Bailey, "40-Hour Work Week: Its History and Future," *actiPLANS* (blog), May 2024, https://www.actiplans.com/blog/40-hour-work-week.

114 Erik Rauch, "Productivity and the Workweek," MIT Computer Science and Artificial Intelligence Laboratory Project on Mathematics and Computation, accessed March 20, 2025, https://groups.csail.mit.edu/mac/users/rauch/worktime/.

115 Keynes, *Essays in Persuasion*, 368.

AI AS A TOOL

Without question, having a purposeful job is essential for our mental well-being.[116] Stable employment, a steady income, and strong social connections are all closely linked to good mental health.[117] Conversely, poverty, financial struggles, and lack of social support are major drivers of mental health issues.[118] Economic downturns affect critical social determinants of health such as employment, education, income, and nutrition, which can lead to increased rates of depression.[119] Research shows that job loss significantly raises the risk of severe depression and even suicide, particularly among young people.[120]

Recognizing that meaningful work is essential to the human psyche, it becomes clear that, like draft horses, humans need labor—not just for survival, but to fulfill a deeper need. With this understanding, we can start to see AI not as a technology

---

116 Ian D. Boreham and Nicola S. Schutte, "The Relationship Between Purpose in Life and Depression and Anxiety: A Meta-Analysis," *Journal of Clinical Psychology* 79, no. 12 (December 2023): 2736–2767, https://doi.org/10.1002/jclp.23576; Angelina R. Sutin et al., "Purpose in Life and Stress: An Individual-Participant Meta-Analysis of 16 Samples," *Journal of Affective Disorders* 345 (January 2024): 378–385, https://doi.org/10.1016/j.jad.2023.10.149.

117 Robert E. Drake and Michael A. Wallach, "Employment Is a Critical Mental Health Intervention," *Epidemiology and Psychiatric Sciences* 29 (2020): e178, https://doi.org/10.1017/S2045796020000906.

118 Peter Warr and Paul Jackson, "Factors Influencing the Psychological Impact of Prolonged Unemployment and of Re-Employment," *Psychological Medicine* 15, no. 4 (1985): 795–807, https://doi.org/10.1017/S0033291700000502X; Tom Fryers et al., "The Distribution of the Common Mental Disorders: Social Inequalities in Europe," *Clinical Practice and Epidemiology in Mental Health* 1, no. 14 (2005), https://doi.org/10.1186/1745-0179-1-14; David Dooley et al., "Depression and Unemployment: Panel Findings from the Epidemiologic Catchment Area Study," *American Journal of Community Psychology* 22, no. 6 (December 1994): 745–765, https://doi.org/10.1007/BF02521557; Mary de Groot et al., "Depression and Poverty Among African American Women at Risk for Type 2 Diabetes," *Annals of Behavioral Medicine* 25, no. 3 (June 2003): 172–181, https://doi.org/10.1207/S15324796ABM2503_03.

119 Astier M. Almedom, "Social Capital and Mental Health: An Interdisciplinary Review of Primary Evidence," *Social Science and Medicine* 61, no. 5 (September 2005): 943–964, https://doi.org/10.1016/j.socscimed.2004.12.025; World Health Organization Regional Office for Europe, "EUR/RC59/7: Health in Times of Global Economic Crisis: Implications for the WHO European Region," paper presented at the fifty-ninth session of the WHO Regional Committee for Europe, Copenhagen, September 14–17, 2009, https://iris.who.int/bitstream/handle/10665/342835/59wd07e-GlobalEconCrisisHealth-90418.pdf?sequence=1&isAllowed=y.

120 Matteo Picchio and Michele Ubaldi, "Unemployment and Health: A Meta-Analysis," *Journal of Economic Surveys* 38, no. 4 (September 2024): 1437–1472, https://doi.org/10.1111/joes.12588.

that simply replaces human effort but as a tool, much like the horse or tractor, that enhances it. AI has the potential to boost creativity and productivity, just as horses once helped humans achieve more efficient farming outcomes.

Just as the adoption of draft horses increased productivity and required new skills—like leatherworking, metalworking, and the building of barns—this new AI era demands a rethinking of work and a significant upskilling in AI-related technologies. By reimagining our roles and how we interact with these advanced tools, we can use AI not as a replacement but as a collaborative partner. This approach blends technology with human potential, forging a future where their combined capabilities open up new possibilities.

Addressing the challenges brought by AI in the workforce will require a major commitment to reskilling and upskilling those displaced by technological advances. Ensuring accessible educational opportunities is crucial, and this will need collaboration between governments, educational institutions, and the private sector. Companies that benefit from AI must also take the lead in supporting these efforts, recognizing their social responsibility.

Ongoing support for career transitions is equally important as the job market continues to evolve. Governments can help by establishing policies that promote continuous learning and adaptation, including tax incentives for companies that invest in workforce reskilling and protections for workers impacted by AI-driven job changes.

Embracing a culture of lifelong learning is key to navigating this transition. Education and skill development must be seen as an ongoing process, crucial for career longevity in an AI-enhanced world. This cultural shift, alongside supportive policies and collaboration, will help mitigate the risks of job displacement caused by AI.

Ultimately, the goal is to use AI as a tool to elevate our intellectual, creative, and productive potential, making our lives more efficient and meaningful. With the right approach, AI can enhance human roles if we foster a partnership that elevates our intellectual and productive potential, rather than displacing us.

The real challenge is making sure we don't lose our sense of purpose along the way.

CHAPTER 9

AI AND GLOBAL INEQUALITY

"The Internet is not a luxury; it is a necessity."

—PRESIDENT BARACK OBAMA, 2015

"I think we are going to drive the cost of Intelligence down so close to zero that it will be this 'before and after' transformation for society."

—SAM ALTMAN, CEO, OPENAI

Somewhere above the Philippine Archipelago, 26,000 feet in the air, I woke up crammed between a crush of weary souls on a military C-130 cargo plane. The air was thick and heavy, filled with the stench of unwashed bodies and the palpable exhaustion shared by both refugees and first-world disaster responders.

At that moment, I was the only one awake. Still groggy, I glanced out the small window and noticed something surreal: a UMV drone gliding effortlessly alongside our plane. Its camera was trained on us, capturing images and video of our cramped conditions. This sleek piece of cutting-edge technology floated with ease in the open sky, a stark contrast to the grim reality inside the aircraft—where every inch of space was filled with the aftermath of human suffering.

Just a week earlier, Super Typhoon Haiyan—known locally as Yolanda—had torn through the Philippines with unprecedented fury. Striking at 20:40 UTC on November 8, 2013, this monstrous force of nature delivered sustained winds of 195 mph and gusts that topped a staggering 235 mph. But it wasn't just the wind that made it deadly. The real weapon was the tidal surge—a 24.6-foot wave that flattened entire towns, swept away buildings, and carried thousands of lives out to sea—for good.

The aftermath defied words. Massive tanker ships lay stranded in city streets, semitrucks filled with coconuts—the economic lifeblood of the region—were embedded in the ruins of homes. Tens of thousands of houses were leveled beyond recognition. Live electrical wires snaked through pools of seawater, hidden death traps waiting for survivors who dared to emerge from the wreckage. The World Health Organization classified the storm as a Category 3 disaster—the highest possible rating, placing it on par with the 2004 Indian Ocean tsunami and the 2010 Haiti earthquake. In the modern era of recorded storms, there was no equal.

The statistics were staggering: 16.1 million people were affected, 4.1 million displaced, 6,300 lives confirmed lost, 1,061 swept away and still missing, and 28,689 injured.[121] The damage was estimated at an overwhelming $13 billion.[122]

Leyte, the hardest-hit island, with 2.1 million people, was devastated. No hospitals remained functional. Lifelines of

---

121 Eduardo D. Del Rosario, "Updates Re the Effects of Typhoon 'YOLANDA' (HAIYAN)," Republic of the Philippines National Disaster Risk Reduction and Management Council, April 17, 2014, https://ndrrmc.gov.ph/attachments/article/1329/Update_on_Effects_Typhoon_YOLANDA_Haiyan_17APR2014.pdf.

122 Kathryn Reid, "From the Field: Typhoon Haiyan: Facts, FAQs, and How to Help," World Vision, October 17, 2023, https://www.worldvision.org/disaster-relief-news-stories/2013-typhoon-haiyan-facts#.

society as we know it—communication, transportation, infrastructure—all gone, as if civilization had been dragged back into the Stone Age in mere hours.

Our team, Mammoth Medical Missions—a nonprofit surgical strike team I helped found a few years earlier—had been bound for a routine mission in Chiapas, Mexico. But the moment we heard about the typhoon, we changed course. Within fifty-two hours of landfall, we became the first international team to reach ground zero. What we found was a landscape of pure destruction. The entire medical system was wiped out. There were no hospitals, no clinics—nothing.

For the next five days, we turned the town hall—one of only two usable buildings in a city of 57,000—into a makeshift hospital. On the mayor's desk, we performed 109 life-saving surgeries under the most primitive conditions: no power, no running water, no roof over our heads. Among those surgeries were eleven births, two of which were emergency C-sections, bringing new life amid the wreckage of nearly 822 dead bodies surrounding us.

Now, as the drone hovered outside, still filming, I couldn't help but reflect on the strange juxtaposition of this moment. Here we were, leaving a disaster zone that seemed to have ripped society back in time, while this piece of first-world ingenuity floated effortlessly alongside, capturing every detail. It was a potent reminder of the glaring inequalities in technological access.

That drone, so advanced and precise, was documenting the aftermath of a disaster that had reduced millions of people to the most basic fight for survival. The contrast couldn't have been more jarring. It highlighted the paradox of our age: While some parts of the world are developing technologies that seem almost miraculous, others struggle just to survive in conditions that feel painfully archaic.

Will AI become a tool for global equity, or will it reinforce existing disparities? In this chapter, we explore how AI holds the power to either bridge or widen the global socioeconomic divide. You'll see how AI has the potential to transform sectors like education, healthcare, agriculture, and disaster preparedness, offering solutions that could uplift millions. But, at the same time, we'll look at the critical issue of the digital divide, where limited access to technology in developing regions risks deepening existing inequalities. Through real-world examples and an analysis of the current landscape, this chapter calls for a proactive, collaborative effort to ensure that AI serves as a tool for global equity, not a force that perpetuates disparity.

The Paradox: AI has the potential to level the playing field but often exacerbates existing inequities between nations and communities.

The Roses: AI access could democratize education and healthcare.

The Thorns: Wealthier nations dominate AI development, leaving others behind.

THE THORNS
THE DIGITAL DIVIDE

The digital divide marks a critical gap between individuals and regions with access to modern information and communication technologies and those without. This disparity includes not just access to the internet, computers, and smartphones, but also the skills and knowledge needed to use them effectively. The impact of the digital divide goes far beyond information access; it deepens economic and educational inequalities worldwide.

AI, in particular, amplifies these global socioeconomic disparities. Access to AI requires significant computational resources and data—assets largely concentrated in wealthier nations with advanced technological infrastructures. As a result, individuals and countries with these resources reap the benefits of AI advancements, while others are left behind, further widening the socioeconomic divide.

This divide isn't solely about devices or internet access; it reflects profound differences in economic opportunity, education, and digital literacy essential for thriving in a tech-driven world. Developed nations possess the infrastructure, educational systems, and economic structures to support cutting-edge technologies like AI. These countries can leverage AI to enhance areas such as healthcare, finance, education, and public safety, creating a cycle of innovation and benefit.

In contrast, developing regions face substantial barriers that limit their access to AI. The gap is staggering, with wealthier countries positioned to lead the AI revolution while others struggle to gain a foothold. How big is the gap?

Alarmingly so.

AI and the Wealth Disparity

According to the Henley & Partners *The USA Wealth Report 2024*, the United States is at the pinnacle globally in terms of liquid investable assets, holding a substantial $67 trillion.When it comes to wealth per capita, the US secures the sixth position worldwide with $201,500, following affluent nations such as Monaco, Luxembourg, Switzerland, Australia, and Singapore.[123]

123　Dominic Volek et al., "The USA Wealth Report 2024," Henley & Partners, accessed March 20, 2025, https://www.henleyglobal.com/publications/usa-wealth-report-2024.

The US boasts a remarkable 5.5 million millionaires, accounting for 37 percent of the global total, along with 9,850 centimillionaires—individuals whose net worth exceeds $100 million.[124] As of April 2024, the US has a record 813 billionaires, the highest number globally.[125] Leading the list of the wealthiest are tech giants like Jeff Bezos of Amazon, Larry Ellison of Oracle, and Mark Zuckerberg of Facebook.[126] At the top of the wealthiest individuals are tech giants like Elon Musk, coming in at number two behind only French luxury goods mogul Bernard Arnault, with an estimated fortune of $195 billion, followed by Jeff Bezos, founder of Amazon, at $194 billion. Mark Zuckerberg, founder of Facebook, holds $177 billion, while Oracle co-founder and chairman Larry Ellison rounds out the top five at $141 billion.[127]

In the technology sector, the *Forbes World's Billionaires* list for the year featured 342 billionaires, up from 313 the previous year. Collectively, these tech billionaires hold over $2.6 trillion—more than any other industry. This represents a remarkable $750 billion increase from 2023, the largest growth across all sectors.[128]

For comparison, let's consider a country familiar to most Americans: Afghanistan. It ranks 164th out of 195 countries on

---

124 Sarah Nicklin, "USA Wealth Report 2024," press release, Henley & Partners, March 19, 2024, https://www.henleyglobal.com/newsroom/press-releases/usa-wealth-report-2024.

125 Devin Sean Martin, "The Countries with the Most Billionaires 2024," *Forbes*, April 2, 2024, https://www.forbes.com/sites/devinseanmartin/2024/04/02/the-countries-with-the-most-billionaires-2024/.

126 Phoebe Liu, "Tech Billionaires Have Added an Astonishing $750 Billion to Their Fortunes over the Past Year," *Forbes*, April 2, 2024, https://www.forbes.com/sites/phoebeliu/2024/04/02/tech-billionaires-have-added-an-astonishing-750-billion-to-their-fortunes-over-the-past-year/.

127 Chase Peterson-Withorn, "Forbes' 38th Annual World's Billionaires List: Facts and Figures 2024," *Forbes*, April 2, 2024, https://www.forbes.com/sites/chasewithorn/2024/04/02/forbes-38th-annual-worlds-billionaires-list-facts-and-figures-2024/.

128 Liu, "Tech Billionaires."

the global wealth index, starkly highlighting its economic struggles. In 2023, Afghanistan's GDP per capita was $415.70—a slight increase from $357.30 in 2022 but a significant drop from its high of $651.40 in 2012. Afghanistan's GDP per capita dropped precipitously in 2021 to $356.50 from $510.80 in 2020.[129] This dramatic decline represents less than 3 percent of the global average GDP per capita, underscoring the severe economic impact of ongoing political turmoil and instability.[130] In stark contrast, the US has an average GDP per capita of approximately $83,000.[131]

This means that for every dollar an Afghan earns, an American earns about $200.

Afghanistan's low GDP per capita is the result of various factors, including political instability, limited access to advanced technology, and low levels of industrialization, a sharp contrast to the high levels of wealth and technological advancement in the US.

In terms of technology and high-tech access, the disparity is equally glaring. The US is a global leader in technology and AI, with widespread internet access and a digitally literate population. This access fuels innovation and economic growth, further strengthening its position in global wealth rankings. Afghanistan, however, struggles with limited access to high-tech resources and AI technologies. Internet penetration is low, at just 18 percent, and the infrastructure needed to support

129 "Per Capita (Current US$)—Afghanistan," World Bank Group, accessed March 20, 2025, https://data.worldbank.org/indicator/NY.GDP.PCAP.CD?locations=AF.

130 Kemp, S. *DIGITAL 2023: Afghanistan*. DataReportal, February 13, 2023.

131 "Per Capita (Current US$)—United States," World Bank Group, accessed March 20, 2025, https://data.worldbank.org/indicator/NY.GDP.PCAP.CD?locations=US.

advanced technology is severely underdeveloped, affecting its position in the global technological and economic landscape.[132]

Access to AI technologies is significantly skewed toward developed countries, which often have the infrastructure, educational systems, and economic frameworks necessary to support advanced technologies. These nations not only use AI directly but also have the resources to innovate and deploy the technology to enhance sectors like healthcare, education, finance, and public safety. Developing countries, on the other hand, face multiple barriers that limit their access to these technologies.

One such barrier is internet speed. The median internet speeds in Kabul are far below global averages, reflecting the region's infrastructure challenges. As of 2025, Afghanistan's fixed broadband speeds average around 4.12 Mbps for downloads and 3.37 Mbps for uploads, compared to Silicon Valley, where download speeds often exceed 100 Mbps due to advanced infrastructure and the concentration of high-tech industries.[133]

Internet accessibility is another barrier. As of early 2023, only about 18 percent of Afghanistan's population had internet access, highlighting a substantial digital divide when compared to Silicon Valley, where internet access is nearly ubiquitous.[134] This disparity is compounded by the absence of a robust digital infrastructure, including data centers, which limits the potential for technological advancement and access to AI in Kabul.

132 Simon Kemp, "Digital 2024: Afghanistan," DataReportal, February 23, 2024, https://datareportal.com/reports/digital-2024-afghanistan.

133 "Afghanistan Median Country Speeds Updated March 2025," Speed Test, accessed April 23, 2025, https://www.speedtest.net/global-index/afghanistan; "Internet Speeds in San Francisco, CA," Real Speeds, last updated June 18, 2024, https://realspeeds.com/california/san-francisco.

134 Yan Liu et al., *Digital Progress and Trends Report* (The World Bank, 2024), 112, https://openknowledge.worldbank.org/server/api/core/bitstreams/95fe55e9-f110-4ba8-933f-e65572e05395/content.

This technological gap also extends to education and gender equality. In the US, where female literacy rates exceed 99 percent, women actively participate in STEM fields. In contrast, Afghanistan's female adult literacy rate remains alarmingly low at just 24.15 percent as of 2024. Meanwhile, male adult literacy in Afghanistan stands at 51.99 percent, highlighting deep-rooted gender disparities and sociopolitical barriers that limit women's access to both education and technology.[135] These factors contribute to Afghanistan's ranking among the countries with the lowest literacy rates globally, further exacerbating the divide in technological advancement and access to opportunities.

These differences highlight a troubling reality: As AI continues to advance, the wealth and innovation it generates could disproportionately benefit technologically advanced regions like Silicon Valley, widening global inequalities. Without concerted efforts to bridge these divides, regions like Afghanistan may become increasingly marginalized in the emerging digital and AI-driven economy.

Double-Edged Sword: AI-Driven Automation & Economic Inequality

Economic disparities between wealthy and poorer regions are being further widened by the rise of AI-driven automation, particularly in how it reshapes job opportunities across global industries. In wealthier nations, industries benefit greatly from AI, which boosts efficiency and drives innovation. In sectors like manufacturing and financial services, AI streamlines operations, optimizes supply chains, and provides personal-

---

135 "Afghanistan Population," CountryMeters, accessed March 20, 2025, https://countrymeters.info/en/Afghanistan.

ized customer experiences, increasing both productivity and profitability.

However, the effects of AI automation on job opportunities in poorer regions can be starkly different. Traditional industries such as agriculture and manufacturing, which employ a large portion of the workforce in these areas, face significant job displacement due to automation. In agriculture, AI technologies like automated harvesting machines and precision farming reduce the need for manual labor, causing job losses for workers who may not have the skills to transition into new roles. Similarly, in manufacturing, automation replaces jobs that involve repetitive tasks, further threatening low-skill employment.

This transition poses a major challenge for regions where educational and training systems are not equipped to help the workforce develop the skills needed for more technologically advanced jobs. As a result, the economic gap continues to widen—countries that can effectively harness AI keep advancing, while those unable to adapt face growing economic stagnation and increasing social challenges.

For poorer regions, the critical issue is how to navigate this transition. Investing in education and skills training is essential, alongside developing policies that encourage technological adaptation while minimizing the negative impact on employment. Without proactive measures, the benefits of AI are likely to remain unevenly distributed, deepening global economic inequalities.

But what if we could change this narrative? What if, instead of reacting after the fact, the global community came together to proactively create AI systems designed specifically to uplift third-world countries, rather than leaving them behind?

THE ROSES
BRIDGING THE DIGITAL DIVIDE

Imagine a world where AI isn't just about generating wealth but about driving real, meaningful change—closing the gap between developed and developing nations, and fostering growth and equality. Instead of focusing solely on the risks, we'd be using AI's vast capabilities to tackle some of humanity's most urgent challenges, distributing its benefits in a way that truly uplifts everyone.

If we can bridge the digital divide, AI has the power to transform sectors like education, healthcare, agriculture, and disaster preparedness—especially in regions currently left behind by modern technology. The potential is enormous: reducing inequalities, improving quality of life, and empowering millions around the world.

Let's take a closer look at what's possible.

In Education

Overcoming the digital divide and harnessing the power of AI has the potential to dramatically transform education in third-world countries. AI can provide personalized learning experiences, tailoring educational content to meet the unique needs of each student. This would be especially impactful in regions with overcrowded classrooms and limited teaching resources, ensuring that education can adapt to every student's pace, style, and learning preferences.

AI-powered platforms offer another promising solution. These online platforms can extend the reach of quality education to even the most remote and underserved communities. With access to a variety of courses—from basic literacy to advanced STEM subjects—students in third-world countries can tap into world-class educational resources. By improving digital network access, AI could help raise literacy rates and broaden educational opportunities for those who have historically been left behind.

And let's not overlook the impact AI could have on women and girls. In many parts of the world, they face significant barriers to education, but AI could provide them with a way forward. Digital learning platforms could be accessible within the home or local community centers, offering safe spaces for girls to learn. By tailoring AI-powered courses to their needs, we're talking about empowering them with skills in financial literacy, health education, and rights awareness. This isn't just about education—it's about elevating their role in society and giving them the tools to create change.

The potential of AI isn't limited to basic education, either—it extends to specialized fields like STEM. By using data analytics, AI can identify students' strengths and interests early on, delivering specialized content to foster their skills in science, technology, engineering, and mathematics. This targeted approach can cultivate a new generation of innovators and problem solvers, crucial for both local development and global progress.

The COVID-19 pandemic demonstrated the effectiveness of remote learning, which can be particularly beneficial in third-world countries where political, geographical, or socioeconomic challenges often disrupt schooling. AI can enhance these platforms by automating feedback, assessing student progress, and facilitating interactive learning experiences that mimic face-to-face education. In this way, AI ensures that students in underdeveloped regions can continue learning, even in the most challenging circumstances.

A powerful example of AI transforming education is the

Can't Wait to Learn program. This initiative provides tablet-based learning to disadvantaged children in conflict-affected regions, including young girls. Through engaging digital games, children access quality educational content designed to be culturally relevant, helping them develop vital skills in mathematics and language.[136]

One standout feature of the program is its use of machine learning to tackle "wheel-spinning," where children become stuck on particular tasks and risk losing motivation. Machine-learning algorithms analyze performance to determine why a child is struggling—whether it's due to forgetting prior lessons or misunderstanding the current material. The system then guides them to revisit essential concepts, keeping them engaged and helping them progress.

By integrating AI in this way, the program not only addresses the lack of traditional schooling but also enhances the efficiency of learning interventions. This personalized support is invaluable in large classrooms or areas with a shortage of professional teachers, ensuring that each child continues to advance at their own pace.

That said, there's still a major hurdle to overcome: the digital divide. In many parts of the world, access to the internet and digital devices is severely limited. Solving this issue requires serious investment in infrastructure and policy efforts to ensure that technology is accessible to everyone. Governments, NGOs, and international organizations need to collaborate to expand digital networks, lower the cost of technology, and ensure that power sources are sustainable and reliable.

---

136 Tong Mu et al., "Towards Suggesting Actionable Interventions for Wheel-Spinning Students," in *Proceedings of The 13th International Conference on Educational Data Mining (EDM 2020)*, eds. Anna N. Rafferty et al. (International Educational Data Mining Society, 2020), 189, https://files.eric.ed.gov/fulltext/ED608061.pdf.

In Healthcare

AI also has the power to completely transform healthcare, especially in third-world countries where access to medical professionals and facilities is severely limited. Imagine what could happen if we overcame the digital divide in this sector. AI could revolutionize healthcare by bringing cutting-edge diagnostics, telemedicine, and resource management to areas that desperately need it. We're talking about drastically improving the overall level of health and making healthcare more accessible to those who need it most.

One of the most exciting possibilities is AI in diagnostics. Right now, AI-powered diagnostic tools can analyze medical images, recognize patterns, and identify diseases faster and more accurately than even the best-trained human doctors. In places where radiologists and specialists are few and far between, AI could fill the gap, offering consistent and reliable diagnostic support. Think about how much earlier diseases like tuberculosis, malaria, and cancer could be detected in developing countries if AI tools were available. These are conditions that are often caught too late for effective treatment, but AI could change that.

Then there's telemedicine. We've seen how useful it is during crises like the COVID-19 pandemic, but AI can take it even further. Imagine AI-assisted telemedicine platforms that can provide preliminary assessments based on a patient's symptoms, guiding them to the right care without needing a doctor to jump in right away. For people living in rural areas who often have to travel long distances just to see a doctor, this could be a game changer. AI could help manage routine cases and flag more serious conditions, making sure limited medical resources are used where they're needed most.

But AI's potential doesn't stop at diagnostics and telemedicine. It can also improve how healthcare systems are managed. AI algorithms can predict patient admissions, helping hospitals manage their staff and resources more efficiently. It can even streamline supply chain logistics, ensuring that medications and equipment are stocked and distributed based on real-time needs. Think about how this could help areas where medical supplies are often scarce or delayed.

One real-world example of AI making a difference in healthcare is the Visualize No Malaria tool developed by PATH. In Zambia, this AI-powered platform helps health workers track malaria outbreaks by analyzing data from various sources. It helps them understand where the disease is spreading and allocate resources accordingly. This kind of strategic planning and targeted intervention is critical in resource-limited settings. The results? Better management of disease, more effective use of healthcare resources, and ultimately, lives saved.[137]

Now, let's be real. As amazing as all this sounds, there are still some big hurdles to overcome. The lack of infrastructure—things like reliable internet, electricity, and even basic healthcare facilities—makes it hard to deploy these AI tools in many parts of the developing world. On top of that, there's the issue of technological literacy. Many of these communities don't have the training needed to use AI tools effectively. Solving these problems will take serious investment—not just in infrastructure but also in education.

This is where collaboration is key. Governments, NGOs, international health organizations, and private companies all need to come together to fund and facilitate the deployment of AI in healthcare. We're talking about developing the infrastructure, training local professionals, and making sure

---

137 "Visualize No Malaria," Path, accessed March 20, 2025, https://www.path.org/visualize-no-malaria/.

that AI systems are not just dropped into these communities but actually integrated into their healthcare practices. It's about building a sustainable system that works for the long haul.

In Sustainable Agriculture

AI could transform agriculture, particularly in third-world countries where food scarcity, inefficient resource management, and outdated farming methods contribute to ongoing hunger and malnutrition. If we could cross the digital divide and bring AI into the hands of farmers in these regions, it could be a complete game changer—boosting productivity, promoting sustainability, and even helping to eliminate starvation. We're talking about AI-powered precision farming that optimizes every aspect of agriculture, from water use to fertilizer application, right down to the timing of planting and harvesting.

These days, precision farming uses data from satellites, drones, and ground sensors to give farmers detailed insights into their fields. This allows for the exact amount of water, fertilizer, and pesticides to be applied where and when they're needed—no more, no less. Imagine the impact this could have in a region where resources are already stretched thin. This could mean more crops, less waste, and less damage to the environment from overuse of chemicals. In countries where agriculture is the backbone of the economy, this kind of efficiency could completely transform local food systems.

One example of what AI can do is happening right now in Germany with the NaLamKI project. This initiative uses AI to help farmers optimize irrigation, fertilization, and pest control through a cloud-based platform that collects data from satellites, drones, and soil sensors.[138] This platform gives farmers a bird's-eye view of their entire operation and helps them make smarter decisions about how to use their resources. And the results? Better crop yields, lower costs, and healthier ecosystems. If we could replicate this kind of technology in third-world countries, we could start addressing some of the biggest challenges they face—like climate change, resource scarcity, and biodiversity loss—all while producing more food.

I've always been struck by how interconnected agriculture and the environment are. When we get it right, farming doesn't just feed us—it preserves our ecosystems and sustains the planet for future generations. AI helps us get it right by enabling more sustainable farming practices. By precisely managing inputs like water and fertilizers, AI reduces runoff into rivers and lakes, prevents soil degradation, and helps preserve the natural ecosystems that are so vital for long-term agricultural productivity.

There's also an economic side to this. AI can automate a lot of the day-to-day farming operations, which reduces labor costs—something that's a big deal in rural areas where there are often workforce shortages. Automation could even help retain younger populations in farming communities, offering a more attractive, tech-forward version of farming that's both profitable and efficient. Plus, AI's predictive capabilities mean farmers can better adapt to changing weather patterns, giving them a head start in dealing with the increasingly unpredictable effects of climate change. The value of having that kind of foresight in agriculture can't be overstated—it's literally the difference between a good harvest and losing a year's worth of work.

But here's the catch, and it's a big one: We're not going to see

---

138 Fraunhofer-Gesellschaft, "Smart Farming: AI Technologies for Sustainable Agriculture," PhysOrg, November 2, 2021, https://phys.org/news/2021-11-smart-farming-ai-technologies-sustainable.html.

these benefits unless we tackle the major infrastructure challenges holding these regions back. There's the obvious stuff, like lack of internet access and reliable electricity, but it goes deeper than that. Many farmers simply don't have the technological literacy to make full use of AI tools. Overcoming these hurdles is going to require serious investments—not just in physical infrastructure, but in education and training. Farmers need to know how to use these tools, how to interpret the data, and how to adapt their practices based on the insights AI provides.

This is where collaboration comes in. Governments, tech companies, NGOs, and farming communities all need to work together to bridge this gap. We need policies that incentivize the adoption of AI in agriculture, and we need partnerships that ensure these technologies are accessible, affordable, and easy to use. It's not just about creating fancy AI tools—it's about making sure those tools reach the farmers who need them most and are integrated into their everyday operations.

In Disaster Preparedness

And finally, AI has the potential to reshape how vulnerable regions manage and respond to natural disasters, particularly in storm prediction and disaster readiness. Right now, access to advanced AI technologies that can predict storm intensities and trajectories is mostly limited to first-world countries. Ironically, many of these nations are less prone to severe natural disasters compared to regions like the Philippines, Iraq, or Nepal, which are hit hard by storms and earthquakes but lack the technology to prepare effectively.

Now imagine if we could bridge that digital divide. Countries regularly hit by disasters could have the same predictive tools as the US or Europe. Take Super Typhoon Haiyan, for example. If AI-driven predictive tech had been available in the Philippines, they could have forecasted the storm's path and intensity more accurately, days earlier. That means better evacuation strategies, more effective disaster management, and potentially thousands of lives saved—not to mention the reduced impact on infrastructure and the economy.

And we're not just talking about predicting storms. AI can take disaster response to a whole new level.[139] It can manage resources, simulate various disaster scenarios to fine-tune response strategies, and even maintain communication networks by predicting outages and adjusting for network loads.[140] Instead of just reacting to disasters, vulnerable countries could shift to a proactive approach—preparing before disaster strikes, not just scrambling afterward.

Having personally been in the thick of international disaster response, I know all too well the challenges that come with these situations. I was at Ground Zero after the September 11 attacks, I helped out during Super Typhoon Haiyan in the Philippines, and I responded to the 2015 earthquake in Nepal.

Let me tell you, rescuing people from collapsed buildings is no easy task.

At Ground Zero, there were constant threats—collapsing structures, asbestos, live electrical wires, even exploding gas lines. I still think about the long-term health risks, like respiratory problems and delayed cancer diagnoses, that many of

---

139 Bernabe Gomez and Usama Kadri, "Numerical Validation of an Effective Slender Fault Source Solution for Past Tsunami Scenarios," *Physics of Fluids* 35, no. 4 (April 2023), https://doi.org/10.1063/5.0144360; Usama Kadri, "Applying AI-Based Models to Predict Tsunamis," UNESCO, last updated May 14, 2024, https://www.unesco.org/en/articles/applying-ai-based-models-predict-tsunamis.

140 Sarah Derouin, "AI Could Help Refine Tsunami Early Warning Systems," *Civil Engineering*, June 22, 2023, https://www.asce.org/publications-and-news/civil-engineering-source/civil-engineering-magazine/article/2023/06/ai-could-help-refine-tsunami-early-warning-systems; Doyle Rice, "Using AI, Researchers Have Created a New Tsunami Warning System," *USA Today*, April 25, 2023, https://www.usatoday.com/story/news/nation/2023/04/25/new-tsunami-warning-system-uses-artificial-intelligence-ai/11715026002/.

us faced because of those conditions. In the Philippines, we were operating in one of the only two buildings left standing in a city of over 54,000—our makeshift hospital was basically on the brink of collapse itself. And in Nepal, navigating through Kathmandu was terrifying—dangling electrical wires, crumbling infrastructure, and in the rural areas, the poorly constructed buildings were a death trap for both survivors and rescue workers.

But here's where things get hopeful. AI-powered platforms are starting to change the game in disaster response. Imagine having the ability to assess the safety of a building before rescue workers like me enter. It's like having a digital guardian angel checking the risks, helping us make informed decisions, and ultimately saving lives.

This isn't some far-off dream, either—it's already happening.

An article by Tate Ryan-Mosley, published on February 20, 2023, in *MIT Technology Review's* newsletter *The Technocrat*, explores the practical use of AI, particularly machine learning, in responding to disasters like the earthquake in Turkey and Syria.[141] One AI project making waves is xView2, a collaboration between the US Department of Defense, Carnegie Mellon University, Microsoft, and UC Berkeley. This system uses machine-learning algorithms and satellite imagery to rapidly assess the damage in disaster-stricken areas.[142] Back in the day, damage assessments could take weeks, which obviously delayed critical aid. With xView2, we're talking about assessments in hours, sometimes minutes. That kind of speed can

---

141 Tate Ryan-Mosley, ""How AI Can Actually Be Helpful in Disaster Response," *The Technocrat, MIT Technology Review*, February 20, 2023, https://www.technologyreview.com/2023/02/20/1068824/ai-actually-helpful-disaster-response-turkey-syria-earthquake/.

142 "How AI Can Help Disaster Response Efforts," IP Access International, November 17, 2023, https://www.ipinternational.net/how-ai-can-help-disaster-recovery-efforts/.

be the difference between life and death. It's not perfect yet—it relies on satellite imagery, which has its limits, especially when trying to assess damage on the sides of buildings. Plus, there's always the challenge of getting traditional first responders to trust and adopt these new tools.

Still, the potential here is enormous. AI technologies like xView2 are poised to become essential in global disaster management. These tools are transforming how we prepare for and recover from natural disasters, and having witnessed firsthand the chaos that comes with unpreparedness, I find a lot of confidence in what's being developed.

Access to these technologies could also foster a more equitable global distribution of resources. International cooperation and investment in technology infrastructure could help bring AI-driven disaster response capabilities to at-risk regions. Governments, NGOs, and tech companies could collaborate to implement technology transfer programs, sharing knowledge and resources to build local expertise in AI and disaster management.

Furthermore, educational and training initiatives could empower local experts in AI and meteorology, ensuring that the communities most affected by natural disasters have the skills to operate and innovate with these technologies.

CLOSING THOUGHTS: EQUAL ACCESS, GLOBAL EFFORT

The potential of AI is undeniable, and its capacity to transform education, healthcare, agriculture, and disaster preparedness could reshape the future for millions. But AI's promise is not guaranteed to reach everyone unless we take deliberate, focused action. Bridging the digital divide is not just about providing technology—it's about building the infrastructure, education, and global partnerships necessary to ensure AI serves as a force for equity rather than deepening the chasm between developed and developing regions.

In education, AI can close learning gaps, empower women and girls, and provide personalized learning experiences, even in the most remote areas. In healthcare, it has the power to save lives through advanced diagnostics, telemedicine, and resource management in regions where medical care is scarce. In agriculture, AI can build sustainable food systems that feed growing populations while preserving the environment. And in disaster preparedness, AI can offer life-saving predictions and response strategies to vulnerable communities facing devastating natural events. These represent only a small window in what is possible.

The real challenge is ensuring that these benefits reach those who need them most. This means investing in infrastructure, creating policies that promote equitable access, and building partnerships that empower local communities with the tools and knowledge to use AI effectively. By addressing the digital divide head-on, we can turn the thorns of inequality into the roses of opportunity, creating a future where AI drives global progress for all.

CHAPTER 10

DEEPFAKES AND AUTHENTICITY

"Propaganda must confine itself to a few points and repeat them over and over."

—ADOLF HITLER, *MEIN KAMPF,* WRITTEN
IN LANDSBERG PRISON, 1925

The power of propaganda to shape public opinion and drive political agendas is as old as civilization itself.

"Fake news" and disinformation have long served as potent tools, often deployed to rally support for wars, silence dissent, and consolidate control. An insightful exploration of this influence can be seen in the *State of Deception: The Power of Nazi Propaganda* exhibit by the Holocaust Memorial Museum, originally in Washington, DC, and later displayed in Brussels. Through artifacts and media, the exhibit reveals how the Nazi party leveraged propaganda to catapult Adolf Hitler into power, enabling them to transform political dominance into a frightening military machine.[143]

Joseph Goebbels, Nazi Germany's Reich Minister of Propaganda, deftly wielded print, film, radio, and public speeches to spread Nazi ideology. His campaigns demonized Jews, minorities, and the marginalized, underscoring the dangerous potential of media manipulation when controlled by authoritarian regimes.

But this chilling use of disinformation was hardly unique to Nazi Germany.

In ancient Rome, Octavian deployed propaganda to turn the tide against his rival, Mark Antony, accusing him of betraying Rome for Cleopatra's influence. This painted Antony as a traitor and rallied the Senate and the public to Octavian's side.[144]

Centuries later, in 1095, Pope Urban II delivered a rousing sermon in Clermont, France, rallying support for military campaigns against Islamic forces. His sermon described exaggerated and fabricated atrocities allegedly committed by Muslims in the Eastern provinces, including desecration of Christian churches, forced circumcision of Christian men, sexual violence against women, and other horrific tortures ending in death. These vivid claims stirred enthusiasm, mobilizing the First Crusade and ultimately leading to the capture of Jerusalem and the creation of the Kingdom of Jerusalem.[145]

---

143 Samantha Koester, "'Fake News' Is Not New: The Nazis Used It Too, Says Holocaust Exhibit," *Reuters*, January 28, 2018, https://www.reuters.com/article/world/fake-news-is-not-new-the-nazis-used-it-too-says-holocaust-exhibit-idUSKBN1FE2OL/.

144 Jesse Sifuentes, "The Propaganda of Octavian and Mark Antony's Civil War," *World History Encyclopedia*, November 20, 2019, https://www.worldhistory.org/article/1474/the-propaganda-of-octavian-and-mark-antonys-civil/.

145 Nicholas J. Cull et al., *Propaganda and Mass Persuasion: A Historical Encyclopedia, 1500 to the Present* (ABC-CLIO, 2003), 23–24; Dana Carleton Munro, "The Speech of Pope Urban II. At Clermont, 1095," *American Historical Review* 11, no. 2 (January 1906): 231–242, https://doi.org/10.1086/ahr/11.2.231.

During the American Revolution, Benjamin Franklin famously fabricated a newspaper report detailing atrocities purportedly committed by Native Americans allied with the British. This story fanned colonial support for independence when it was needed most in 1782, with thirty-five American newspapers republishing Franklin's fabrication as fact. It wasn't until 1854—long after the Revolution—that the Trenton, New Jersey, *State Gazette* finally revealed the report as a hoax.[146]

In World War I, British authorities spread false stories of German soldiers committing barbaric acts, galvanizing public support, recruits, and funding for the war effort.[147]

The Cold War took the use of fake news to new heights as both the Eastern and Western blocs deployed disinformation in psychological warfare. Joseph Stalin, known for coining the term *desinformatsiya* (disinformation), wielded it as a tool to maintain his power, silence dissent, and promote Soviet ideological superiority. Stalin's propaganda regime rewrote history, erased political rivals from public memory, and projected an image of strength to the West.[148] According to American historian Timothy D. Snyder, Stalin's policies led to the deaths of around six million people, rising to nine million when accounting for deaths from predictable outcomes of these policies.[149] Another historian, William D. Rubinstein, estimated Stalin's responsibility for at least seven million deaths[150]—approxi-

---

146 Rebecca Onion, "The Atrocity Propaganda Ben Franklin Circulated to Sway Public Opinion in America's Favor," *Slate*, July 1, 2015, https://slate.com/human-interest/2015/07/history-of-benjamin-franklin-diplomacy-propaganda-newspaper-with-stories-of-native-american-atrocities.html.

147 Cull et al., *Propaganda and Mass Persuasion*, 25.

148 Ronald L. Mendell, *Investigating Information-Based Crimes: A Guide for Investigators on Crimes Against Persons Related to the Theft or Manipulation of Information Assets* (Charles C Thomas, 2013), 45–58.

149 Timothy Snyder, *Bloodlands: Europe Between Hitler and Stalin* (Basic Books, 2010), 384.

150 William D. Rubinstein, *Genocide* (Routledge, 2004), 211.

mately 4.2 percent of the Soviet Union's population. For Stalin, propaganda was a weapon as powerful as any military force.

More recently, propaganda facilitated one of the most tragic genocides of the modern era. Leading up to and during the 1994 Rwandan genocide, radio stations and newspapers filled the airwaves with hate speech against the Tutsi population. The infamous radio station RTLM broadcast inflammatory rhetoric that fueled the violence, with fatal consequences. Between April and July 1994, more than 800,000 Tutsis and moderate Hutus were killed in Rwanda. Additionally, around 250,000 women endured sexual violence, many of whom were subsequently killed, and 70 percent of those who survived were infected with HIV. By the end of the hundred-day massacre, 85 percent of the Tutsi population—representing 10 percent of Rwanda's total population—had been exterminated. Half of the country's population was either internally displaced or had fled.[151]

These historical snapshots underscore the staggering power of propaganda to incite violence, sway minds, and reshape societies. As we grapple with a new wave of digital disinformation and "fake news," understanding the deep roots of these tactics reminds us of the urgent need for vigilance.

This historical backdrop sets the stage for examining the modern evolution of propaganda: digital disinformation, bots, and deepfakes. Today, as artificial intelligence and other digital forgeries blur the line between truth and illusion, the lessons of history urge us to look beyond the surface, questioning who controls the message—and to what end.

Can we embrace the creative potential of AI while safe-

---

151 Matthew Lower and Thomas Hauschildt, "The Media as a Tool of War: Propaganda in the Rwandan Genocide," *Human Rights and Conflict Resolution* 2, no. 1 (May 9, 2014): 1, http://www.hscentre.org/wp-content/uploads/2014/05/HRCR-2014-Issue-2-No.-1.-Lower-and-Hauschildt-The-Media-as-a-Tool-of-War.pdf.

guarding against its capacity for deception? In this chapter, we'll explore the roses and thorns of deepfake technology, delving into both the remarkable potential and the profound ethical challenges AI presents. On one hand, deepfakes showcase human ingenuity and promise transformative applications in fields like art and education, pushing the boundaries of creative expression and immersive learning. On the other, they pose significant risks as tools of misinformation and manipulation, capable of undermining trust and destabilizing societies. Through examining both the benefits and dangers of AI's development, this chapter emphasizes the importance of fostering innovation with a balanced approach, advocating for robust ethical standards, regulatory safeguards, and public education to help ensure these powerful technologies are used responsibly and contribute positively to society.

The Paradox: AI enables new forms of creative expression but also threatens truth and authenticity.

The Roses: Creative AI tools empower artists and storytellers.

The Thorns: Deepfakes erode trust in media, creating opportunities for misinformation.

But first, let's set the record straight when it comes to spreading misinformation: We as humans love doing it.

MISINFORMATION: IT'S NOT JUST THE BOTS

While bots play a role in spreading falsehoods, humans remain at the heart of the misinformation cycle.

Rachel Xu of the Yale Jackson School for Global Affairs

explores the environments conducive to misinformation, highlighting the challenges in data protection and the profound human impact on the spread of false information. The digital landscape today is ripe for misinformation, fueled not solely by bots or malicious actors but by our own human interactions within these systems.[152]

Despite the role of bots, it's often human behavior and interaction dynamics within these platforms that drive misinformation's spread. False facts alone may seem harmless; the real danger unfolds as these fragments interact with individual biases, translating into real-world consequences. A striking example comes from a 2018 MIT study analyzing stories shared between 2006 and 2017, which revealed that false information travels six times faster than the truth to 1,500 people. Put simply, it takes true stories six times as long to reach an audience of 1,500 as it does for false stories.[153] Interestingly, this rapid spread isn't driven by bots—both true and false news spreads at similar rates through automated systems. Instead, it's *humans* who, drawn to novelty and emotional weight, amplify falsehoods more readily.[154]

The MIT study highlights several metrics to quantify the spread of misinformation: False news stories are 70 percent more likely to be retweeted than true stories. On Twitter, sequences of unbroken retweet chains, known as "cascades," see falsehoods reaching a depth of ten, about twenty times faster

---

152 Rachel Xu, "You Can't Handle the Truth: Misinformation and Humanitarian Action," *Humanitarian Law and Policy Blog*, International Committee of the Red Cross, January 15, 2021, https://blogs.icrc.org/law-and-policy/2021/01/15/misinformation-humanitarian/.

153 Soroush Vosoughi et al., "The Spread of True and False News Online," *Science* 359, no. 6380 (2018): 1148, https://doi.org/10.1126/science.aap9559.

154 Peter Dizikes, "Study: On Twitter, False News Travels Faster Than True Stories," MIT News, March 8, 2018, https://news.mit.edu/2018/study-twitter-false-news-travels-faster-true-stories-0308.

than factual information. These false cascades often contain surprising or anger-inducing information, which captures attention and spurs rapid sharing.

The sheer volume of information today further complicates efforts to verify news. Traditional fact-checking and editorial standards can barely keep pace with the constant torrent of online content, where anyone can publish without adhering to journalistic standards. This overload leads individuals to rely more on their preexisting biases, preferring information that confirms their viewpoints, thus perpetuating a cycle of misinformation.

This cycle becomes particularly vicious during crises like the COVID-19 pandemic, where misinformation thrives on heightened fear and bias. Health crises can amplify rumors and conspiracy theories, leading to stigma and even violence against certain groups. Increased dependence on digital tools for public health data during these times can also make people more vulnerable to privacy compromises. In crisis-driven scenarios, people may sacrifice data privacy for perceived security, opening doors for scams and unauthorized data collection.

The complexities of today's information landscape underline the urgent need to foster digital literacy and critical engagement. As we navigate this flood of information, these skills are essential to disentangle fact from fiction and mitigate the societal risks that misinformation poses.

HELLO, DEEPFAKES

In case you are unfamiliar, deepfakes—a fusion of "deep learning" and "fake"—rely on sophisticated artificial intelligence and machine-learning techniques to create highly realistic, often undetectable, photo, video, and audio fabrications. These digital forgeries often involve swapping one person's face or voice with another's in a way that is nearly indistinguishable from real footage, making it appear as though someone is saying or doing something they never actually did.

The backbone of deepfake technology lies in generative adversarial networks (GANs). In this process, two machine-learning models work in constant opposition: One generates synthetic images or videos, while the other tries to identify them as fake. This back-and-forth continues until the detector can no longer tell the difference, resulting in a highly convincing fake that is virtually indistinguishable from real footage.[155]

With the rise of digital technologies, the age-old tactics of disinformation have transformed, becoming more pervasive and harder to detect. The digital era has amplified the speed and reach of misinformation, enabling both state and nonstate actors to spread false narratives at a scale previously unimaginable.

Today, tools like deepfakes and other forms of digital manipulation go beyond mere misinformation; they reshape perceptions of reality itself. This technological shift force-multiplies the tactics once wielded by figures like Goebbels and Stalin, scattering their influence across the digital landscape and posing fresh challenges to societies as they grapple with the ever-thinning line between truth and fabrication.

But it's certainly not all bad.

The rise of deepfakes sheds light on this chapter's paradox. While yes, deepfakes can be weaponized for misinformation, used to infringe on personal privacy, or deployed to incite vio-

---

155   Scott Robinson et al., "What Is a Generative Adversarial Network (GAN)?," TechTarget, last updated October 2024, https://www.techtarget.com/searchenterpriseai/definition/generative-adversarial-network-GAN.

lence and unlawfully influence public opinion, they also have transformative capabilities, from animating historical figures for educational purposes to enhancing realism in virtual environments, even allowing deceased artists to "perform" once more. These possibilities underscore the power of AI to create, inspire, and entertain.

Deepfakes thus encapsulate both the "roses and thorns" of AI, which we'll explore next.

THE ROSES
DEEPFAKE TECHNOLOGY

Deepfakes initially gained attention for their applications in entertainment and satire, showcasing the potential of the technology. They allowed content creators to produce compelling videos featuring well-known figures in scenarios they never actually participated in, demonstrating both the creative and technical capabilities of deepfakes. This early use highlighted the technology's ability to enhance storytelling by bringing imaginative scenarios to life, contributing to new forms of artistic expression.

Media Production: A New Era in Cinema

Deepfakes are reshaping media production in ways previously confined to science fiction.

Hollywood's use of advanced visual effects in *Rogue One: A Star Wars Story* is a prime example. Filmmakers utilized advanced visual effects technology to resurrect the characters of Grand Moff Tarkin and Princess Leia, originally portrayed by Peter Cushing and Carrie Fisher, respectively. Peter Cushing, who passed away in 1994, was brought back to life on screen using a combination of CGI and deepfake technology to superimpose his younger facial features onto an actor bearing a similar physique. This technique allowed the character to be seamlessly integrated into the film, which is set just before the events of the original 1977 *Star Wars* movie.[156]

Similarly, Carrie Fisher's Princess Leia was digitally re-created to appear as she did in her youth for the final scene of *Rogue One*. This was achieved by using digital effects to de-age Fisher's appearance from earlier footage, along with a body double for physical scenes. The effect aimed to bridge the gap between *Rogue One* and the original trilogy, providing a visual continuity that fans would recognize.

While these digital resurrections stirred excitement over the ability to honor past performances, they also sparked ethical debates around using actors' likenesses posthumously. Audiences were divided—some praised the technology's capabilities, while others found the effects uncanny or unsettling. These portrayals, however, underscore deepfakes' evolving role in cinema, demonstrating both their storytelling potential and the ethical questions they raise.

Authentic and Synthetic Celebrities

The music industry has also benefited from deepfake technology, notably in reviving the voices of iconic artists.

Country star Randy Travis, who suffered a stroke in 2013 that left him unable to sing, released new music thanks to AI deepfakes. After his stroke, Travis experienced severe aphasia, limiting his ability to speak and sing. However, with the help

---

156 Corey Chichizola, "*Rogue One* Deepfake Makes Star Wars' Leia And Grand Moff Tarkin Look Even More Lifelike," Cinemablend, December 9, 2020, https://www.cinemablend.com/news/2559935/rogue-one-deepfake-makes-star-wars-leia-and-grand-moff-tarkin-look-even-more-lifelike.

of AI, Travis's voice has been digitally reconstructed, allowing him to sing again.

Warner Music Nashville collaborated with London-based developers to create a proprietary AI model. This model was trained on forty-two samples from Travis's extensive discography. By overlaying these AI-generated vocals onto a base provided by singer James Dupré, they were able to re-create Travis's iconic voice. The result was a new song titled "Where That Came From," which captures the essence of Travis's unique sound and style. This incredible use of AI allowed Travis to continue his musical career despite his physical limitations. It also represents a significant technological advancement in how AI can be used to assist artists in overcoming challenges. Travis's wife, Mary, expressed hope that this technology might set a new standard for using AI in music, highlighting its potential to breathe new life into the careers of artists facing similar obstacles.[157]

On the frontier of synthetic celebrities, figures like Lil Miquela—an entirely AI-generated influencer—blur the line between reality and digital creation. Miquela, a Brazilian American model, musician, and social media star, has amassed over a million followers and collaborates with brands, engages in activism, and releases music. But she's not a real person. Her presence as an "influencer" demonstrates how deepfake and AI technology can create new types of celebrity, ones that exist entirely in the digital realm but engage with real-world issues and industries, from fashion to social activism.[158]

---

157 Maria Sherman, "With Help from AI, Randy Travis Got His Voice Back. Here's How His First Song Post-Stroke Came to Be," AP News, May 6, 2024, https://apnews.com/article/randy-travis-artificial-intelligence-song-voice-589a8c142f70ed8ccf53af6d32c662dc.

158 Matt Klein, "The Problematic Fakery of Lil Miquela Explained—An Exploration of Virtual Influencers and Realness," Forbes, November 17, 2020, https://www.forbes.com/sites/mattklein/2020/11/17/the-problematic-fakery-of-lil-miquela-explained-an-exploration-of-virtual-influencers-and-realness/.

EDUCATIONAL TRANSFORMATION
THROUGH DEEPFAKES

Deepfake technology holds transformative potential for educational content, offering innovative ways to make learning more interactive and immersive. Through the use of deepfakes, educators can bring historical figures and events to life, providing students with a dynamic and engaging visual and auditory experience. This can be particularly effective in subjects like history, where understanding the context and personalities of the past can greatly enhance learning.

One practical application of deepfakes in education could involve re-creating speeches or debates with historical figures.[159] For example, students could witness a reenacted debate between historical figures such as Winston Churchill and Franklin D. Roosevelt, offering a vivid understanding of their rhetorical skills and political arguments during World War II.

Deepfakes can also be integrated into virtual reality (VR) platforms to create fully immersive educational experiences. For instance, a VR simulation could allow students to "meet" and interact with figures like Marie Curie—the first woman to win a Nobel Prize for her groundbreaking work in physics, chemistry, and the study of radioactivity—or Martin Luther King Jr., engaging in simulated conversations that are informed by historical records but animated through AI. This would enable students to explore historical scenarios or conduct virtual "interviews" with these figures, deepening their engagement with the material.

The MIT project In Event of Moon Disaster is a remark-

159 Nir Eisikovits, "The Slippery Slope of Using AI and Deepfakes to Bring History to Life," *The Conversation*, November 2, 2021, https://theconversation.com/the-slippery-slope-of-using-ai-and-deepfakes-to-bring-history-to-life-166464.

able educational initiative that utilizes deepfake technology to illustrate what might have happened if the 1969 Apollo 11 mission had failed.[160] This project features a realistically re-created video of President Nixon delivering a contingency speech that was prepared in case the astronauts were unable to return home. Although this speech was never actually delivered, the project brings it to life, showing Nixon announcing the tragic loss of Neil Armstrong and Buzz Aldrin.

Codirectors Fran Panetta and Halsey Burgund employed advanced deepfake techniques to achieve this, including deceptive editing, the use of deep learning algorithms to generate a synthetic voice of Nixon, and dialogue replacement technologies to synchronize voice and mouth movements accurately. In Event of Moon Disaster not only explores the historical "what-if" scenario but also serves as a poignant exploration of the potential impacts of misinformation and deepfake technology in modern society.

This project is supported by educational grants at MIT and aims to foster discussions on media literacy, particularly highlighting the importance of discerning real from manipulated media. It provides a powerful educational tool that encourages students to consider the implications of artificial intelligence and synthetic media, making it a significant contribution to discussions about technology's role in shaping public perception and history.

Additionally, the MIT Center for Advanced Virtuality offers a free online course titled *Media Literacy in the Age of Deepfakes*, teaching students to discern between real and manipulated media. With deepfakes woven into online modules simulating

---

160 Francesca Panetta and Halsey Burgund, "In the Event of Moon Disaster," MIT Docubase, accessed March 20, 2025, https://docubase.mit.edu/project/in-event-of-moon-disaster/.

historical events and figures, this course fosters critical thinking around technology's role in shaping perceptions of truth and history.[161]

Together, these educational initiatives illustrate deepfakes' potential to revolutionize traditional learning, offering immersive environments that make historical and contemporary subjects come alive. Yet, they also underscore the importance of cultivating media literacy, ensuring that students are equipped to navigate a world where synthetic media continues to shape perceptions of truth and history.

THE THORNS
DEEPFAKE TECHNOLOGY

While deepfakes showcase immense potential for creativity and education, their darker applications reveal urgent ethical challenges. These powerful technologies demand strict ethical guidelines to prevent misuse, as the digital world's amplification can transform minor issues into major crises. Questions surrounding ownership, intellectual property, and likeness are particularly pressing in this landscape, where boundaries of authenticity blur in unsettling ways.

Consider the viral phenomenon sparked by a single AI-generated song that appeared to feature Drake and the Weeknd. Uploaded by an anonymous TikTok user under the handle @ghostwriter977, the song amassed six hundred thousand streams on Spotify before it was taken down by Universal Music Group, the music label that represents both Drake and

---

161 Stefanie Koperniak, "Fostering Media Literacy in the Age of Deepfakes," MIT News, February 17, 2022, https://news.mit.edu/2022/fostering-media-literacy-age-deepfakes-0217.

the Weeknd.[162] Despite being credited to "Ghostwriter," the track featured voices that eerily resembled the two artists. Music journalist Jacques Morel voiced concern in a CBS Toronto interview, pointing out that we may soon face a world where it's increasingly difficult to distinguish deepfakes from reality.[163] Such misuse of an artist's likeness could lead to unwanted associations with content they never endorsed.

This incident raises more questions than answers, reminding us that we're at the very beginning—not the end—of this conversation: What protections do artists have in a digital age where their voices, images, and work can be so easily replicated? How can creators safeguard their unique contributions from unauthorized use? And as technology advances, what rights should artists hold over digital copies of their likeness?

The implications of deepfake technology, however, extend far beyond the realms of entertainment and music videos. The growing prevalence of deepfake technology represents a significant challenge in the realm of information integrity, particularly in political contexts.

Political Manipulation

Revisiting the historical tactics of figures like Goebbels and Stalin, and the use of media in the Rwandan genocide, we see that these core issues—propaganda, influence, and manipulation—are as relevant today as ever, only magnified by

---

162 Samantha Murphy Kelly, "The Viral New 'Drake' and 'Weeknd' Song Is Not What It Seems," CNN, April 19, 2023, https://www.cnn.com/2023/04/19/tech/heart-on-sleeve-ai-drake-weeknd/index.html; Faustine Ngila, "Spotify and Apple Music Removed an AI-Generated Fake Song by Drake and The Weeknd," Quartz, April 18, 2023, https://qz.com/drake-the-weeknd-ai-song-spotify-apple-music-1850347902.

163 Jackson Weaver, "Drake and The Weeknd Are Just the Latest Stop on the AI Art Express," CBC News, April 21, 2023, https://www.cbc.ca/news/entertainment/ai-music-drake-weeknd-1.6819092.

technology's reach. Deepfakes could profoundly distort political landscapes, with the potential to manipulate public opinion, disrupt democratic processes, and sow mistrust on a scale once unimaginable.

As nations around the world hold elections, while conflicts and protests escalate, there is a heightened risk that deepfakes will be used to influence public sentiment and undermine democracy.[164] In the ongoing conflict between Ukraine and Russia, for example, media manipulation has become a staple in shaping public perception and international opinion. Both sides employ propaganda and digital misinformation to spin the narrative, but the advent of deepfake technology amplifies these tactics. With the ability to create hyperrealistic videos that falsely depict actions or statements by key figures, deepfakes present a grave challenge to the integrity of information in conflict zones.

A striking example occurred when a deepfake video of Ukrainian President Volodymyr Zelenskyy emerged on social media and briefly appeared on a Ukrainian news site before being taken down. The video showed a rendered likeness of Zelenskyy calling on Ukrainian troops to surrender to Russian forces. Although it was removed swiftly, its potential impact on public morale and perception underscores the power of deepfakes as psychological weapons. Ukrainian officials have suggested that this could be part of Russia's information warfare strategy, warning that such manipulative media could be used to sow fear and confusion.

The video, which lasted about a minute, featured Zelenskyy's likeness with reasonably convincing lip-syncing, but

---

164 "How Disinformation Works—And How to Counter It," *The Economist*, May 2, 2024, https://www.economist.com/leaders/2024/05/02/how-disinformation-works-and-how-to-counter-it.

discerning viewers noticed discrepancies in his accent and slight misalignments between voice and head movements. Hany Farid, a professor specializing in digital forensics at UC Berkeley, remarked, "This is the first one we've seen that really got some legs, but I suspect it's the tip of the iceberg." The incident highlights the growing need for increased vigilance and advanced verification processes to protect against such deceptions. Although major platforms like Facebook, YouTube, and Twitter swiftly removed the video, it continued to circulate on Russian social media, where it gained traction and further amplified false narratives.[165]

The risks extend beyond misleading videos alone; the potential for deepfakes to create widespread confusion in crisis situations is significant. In any conflict, the ability to trust official statements is paramount. However, deepfakes erode this trust, complicating diplomatic efforts and crisis management. The rapid spread of misinformation, coupled with the sheer volume of fabricated content, often overwhelms fact-checkers and AI detection tools, making it nearly impossible to identify and debunk falsehoods before they do real damage.

Deepfakes in the Israel–Palestine conflict illustrate similar risks. In a highly sensitive region where misinformation can have catastrophic consequences, the misuse of deepfake technology could escalate tensions dramatically. By creating hyperrealistic but entirely fabricated portrayals of key figures or events, deepfakes could provoke violent responses or mislead diplomatic efforts based on distorted versions of reality.

Jean-Claude Goldenstein, CEO of CREOpoint—a technol-

165   Bobby Allyn, "Deepfake Video of Zelenskyy Could Be 'Tip of the Iceberg' in Info War, Experts Warn," NPR, March 16, 2022, https://www.npr.org/2022/03/16/1087062648/ deepfake-video-zelenskyy-experts-war-manipulation-ukraine-russia.

ogy company based in San Francisco and Paris that specializes in AI-driven validation of online claims—has warned about the mounting challenges deepfake technology presents. "It's going to get worse—a lot worse—before it gets better," Goldenstein remarked, emphasizing the rise of sophisticated generative AI capable of creating disturbingly realistic images, videos, and audio.[166] CREOpoint has assembled a database cataloging the most viral deepfakes from the Gaza conflict, illustrating the scale of the issue. Such deepfakes can severely misrepresent events, fabricating attacks that never occurred or drastically distorting real incidents. Misrepresentations like these could incite international backlash, strain diplomatic relationships, and even prompt unwarranted international interventions based on false premises.

In the United States, where election integrity is a central focus, deepfakes pose a potent new threat.[167] The technology allows for the creation of highly realistic videos showing politicians making statements or taking actions that never occurred. Imagine if just months before the November 2024 election, a deepfake video had appeared showing President Biden stumbling in confusion during an interview, or one featuring former President Donald Trump angrily calling for protests against the then-current administration. Such fabrications could have created real controversies, casting doubt over candidates right before voting, with fact-checkers racing against the clock to assess authenticity. Minor manipulations can already sway

166 David Klepper, "Fake Babies, Real Horror: Deepfakes from the Gaza War Increase Fears About AI's Power to Mislead," AP News, November 28, 2023, https://apnews.com/article/artificial-intelligence-hamas-israel-misinformation-ai-gaza-a1bb303b637ffbbb9cbc3aa1e000db47.

167 Brandy Zadrozny, "Disinformation Poses an Unprecedented Threat in 2024—And the U.S. Is Less Ready Than Ever," NBC News, January 18, 2024, https://www.nbcnews.com/tech/misinformation/disinformation-unprecedented-threat-2024-election-rcna134290.

public sentiment, but deepfakes elevate this risk by making false portrayals nearly indistinguishable from reality.

Deepfakes could show candidates endorsing or denouncing causes they haven't spoken about, or could even depict fabricated crises meant to mislead voters. With social media as a distribution platform, deepfakes could go viral before being debunked, allowing misinformation to reach and influence millions in critical preelection moments. Without swift verification, such deepfakes could skew public opinion and disrupt election dynamics.

Social Implications: Eroding Trust in Media and Information

Beyond the immediate threat of misinformation, the misuse of deepfake technology poses broader social risks—particularly concerning public trust in the media.

As video and audio content can now be altered with high precision, public confidence in digital information wavers. As skepticism grows, the foundational trust required for healthy democratic discourse and dialogue weakens. This erosion of trust affects not just political news but journalism, public safety announcements, and other vital sources of information. With skepticism rising, democratic discourse and public trust weaken, undermining essential societal functions and creating a culture of doubt.

With all these challenges, it's clear that both government and private organizations need to step up with technologies and strategies to catch deepfake content. Public awareness campaigns can make a huge difference too, helping people recognize the risks of deep fakes and think twice before believing everything they see. Given the global nature of digital disinformation—especially in tense areas like Ukraine and Russia—it may even require international cooperation to establish guidelines and regulations that address these cross-border issues. To really tackle the problem, we need strong measures in place to track and regulate how deepfakes are created and spread. Boosting media literacy, advancing tech solutions for spotting deepfakes, and rolling out clear legal frameworks will all be crucial to minimizing the risks this technology brings. Without these efforts, deepfakes could pose a real threat to the stability and integrity of democracies around the world.

CLOSING THOUGHTS: A QUEST FOR AUTHENTICITY

As we've seen, deepfake technology stands at the heart of a digital crossroads, embodying both the roses and thorns of modern artificial intelligence. Just as propaganda was once wielded to manipulate narratives—from ancient Rome to Nazi Germany to the Rwandan genocide—today's deepfakes carry that same power, amplified by the speed and reach of the digital world. Yet, deepfakes also offer exciting possibilities: they're transforming storytelling, reanimating history for educational purposes, and even giving artists a voice they thought they'd lost. These roses demonstrate how, when used responsibly, deepfakes can inspire, educate, and connect us in ways we couldn't have imagined.

However, the stakes are high. Without thoughtful regulations and widespread media literacy, deepfakes could destabilize trust, blur reality, and undermine democratic processes worldwide. The quest for authenticity in this era of digital forgeries requires vigilance, ethical standards, and a balanced approach—one that embraces innovation while fiercely protecting the integrity of the information we rely on and the society we share.

CHAPTER 11

AI AND THE PARADOX OF WAR

"And Fire Came Down From God out of Heaven and Devoured Them."

—REVELATION 20:9, NEW KING JAMES BIBLE

In the shadowed corridors of history, there's a story of a queen who, on the eve of her marriage to the king, dreamt of a divine thunderbolt striking her womb, igniting a flame that spread "far and wide" before fading to nothing.[168]

In the months that followed, the king himself had a similar vision. He dreamed of sealing the queen's womb with a lion-emblazoned seal—a powerful symbol of royal protection.[169] Together, these secret visions foretold the coming of a child-king destined for greatness, someone whose legacy would echo through the ages.

---

168 Plutarch, *The Age of Alexander: Nine Greek Lives by Plutarch*, trans. Ian Scott-Kilvert (Penguin Books, 1977), 253.

169 Plutarch, *Age of Alexander*, 253.

At the tender age of twenty, after studying under Aristotle, the young prince took the throne and commanded one of the ancient world's most formidable armies. Under his father's rule, the Kingdom of Macedon had risen from a second-rate power to a major force. Now, with this boy-king at its head, Macedon stood poised for unparalleled conquests and influence.

Known to those close to him as Alexandros, he would come down through history as Alexander the Great.

Famed for his tactical genius, Alexander introduced revolutionary military strategies, like the sarissa pike and the tightly organized phalanx formation, which transformed infantry battles.[170] His use of siege technologies like catapults secured his victories and laid the groundwork for future warfare advances, from missile development to space exploration.

But Alexander's impact wasn't limited to the battlefield. His vast empire, stretching from Greece to the edges of India, became a cultural crossroads where Hellenistic and local traditions blended. This fusion sparked an exchange of ideas, arts, and philosophies, creating a period of rich cultural growth that resonated throughout the known world.[171]

Today, echoes of Alexander's strategies even appear in modern business, where tactics derived from his military playbook—like organizational maneuvers and flanking strategies—help companies compete.[172] By spreading Hellenistic

---

170 Colette Hemingway and Séan Hemingway, "The Rise of Macedon and the Conquests of Alexander the Great," Metropolitan Museum of Art *Timeline of Art History* Essay Series, October 1, 2004, https://www.metmuseum.org/essays/the-rise-of-macedonia-and-the-conquests-of-alexander-the-great.

171 *Britannica*, "Alexander the Great's Achievements," accessed March 20, 2025, https://www.britannica.com/summary/Alexander-the-Great.

172 Leonardo Sánchez, "Business Strategy, A Lesson From Alexander The Great," LinkedIn, September 21, 2018, https://www.linkedin.com/pulse/business-strategy-lesson-from-alexander-great-leonardo-s%C3%A1nchez/.

culture, he fueled advancements in science, literature, and the arts, leaving a mark on global culture that endures to this day.

The story of Alexander the Great captures the ancient roots of military innovation—a tradition that has spurred both necessity and ingenuity through the ages. Under his rule, strategic thinking and tactics became an art, setting standards for generations. This legacy of innovation, born from the crucible of war, has had a lasting impact on society, extending far beyond the battlefield.

Just as Alexander's strategies reshaped the ancient world, military innovation has continued to drive breakthroughs that ripple through society, transforming civilian life in ways we often take for granted.

Throughout history, the military has been a powerful engine of innovation, sparking advancements that eventually shape civilian life. This pattern can be traced back through centuries of conflict, where necessities on the battlefield have catalyzed innovations that have reshaped aspects of everyday modern living.

Take the ancient engineering feats of the Roman military, for example. Their roads and bridges, initially built to move armies swiftly, laid the groundwork for modern infrastructure, helping define the field of civil engineering. Or consider the mysterious Greek fire, a naval weapon whose incendiary power inspired early emergency response techniques. Soldiers tasked with handling Greek fire had to be extensively trained in its safe storage, handling, and deployment. This careful management of the volatile substance not only ensured its effective use but also laid the foundation for modern firefighting and hazmat protocols.[173]

---

173 J. R. Parington, *A History of Greek Fire and Gunpowder* (The Johns Hopkins University Press, 1999), 1–4.

Fast-forward to the Napoleonic Wars, which gave rise to canned food—a way to preserve rations for troops that soon revolutionized food storage for everyone.[174] Similarly, during the American Civil War, the military's strategic use of railroads for troop transport set a blueprint for the modern railway systems that fueled national economies.[175]

The pace of innovation only intensified in the twentieth century. World Wars I and II pushed medicine forward, with advances in orthopedic surgery and trauma care developed for wounded soldiers becoming essential pillars of modern healthcare.[176] Radar, initially used to detect enemy aircraft, now supports global air traffic control and weather forecasting.[177]

The trend continued through the Korean and Vietnam Wars, where helicopters were refined for medical evacuations and eventually became vital tools in civilian air rescues and news reporting.[178] More recently, the Gulf Wars and ongoing conflicts brought military technologies like GPS and drones into civilian life, powering everything from navigation systems to environmental monitoring.[179]

---

174 Jerry James Stone, "How Canning Was Invented, and How It Changed the Way We Eat," The Kitchn, April 14, 2015, https://www.thekitchn.com/breakthroughs-in-food-science-canning-218083.

175 Irby W. Bryan, *Civil War Railroads: A Revolution in Mobility* (U.S. Army War College, 2001), 16–19.

176 John Hedley-Whyte and Debra R Milamed, "Orthopaedic Surgery in World War II: Military and Medical Role of Northern Ireland," *Ulster Medical Journal* 85, no. 3 (September 2016): 196–202, https://pubmed.ncbi.nlm.nih.gov/27698524/.

177 "The History of Radar," *Spartan College of Aeronautics and Technology* (blog), November 15, 2021, https://www.spartan.edu/news/the-history-of-radar/.

178 "Medevac," National Museum of the United States Army, accessed March 20, 2025, https://www.thenmusa.org/armyinnovations/innovationsmedevac/; Andrew Pearson, "How Vietnam Changed Journalism," *New York Times*, March 29, 2018, https://www.nytimes.com/2018/03/29/opinion/vietnam-war-journalism.html.

179 Space and Missile Systems Center and SMC History Office, "Evolution of GPS: From Desert Storm to Today's Users," U.S. Air Force, March 24, 2016, https://www.af.mil/News/Article-Display/Article/703894/evolution-of-gps-from-desert-storm-to-todays-users/; Norman Polmar, "Historic Aircraft—The Pioneering Pioneer," *Naval History* 27, no. 5 (September 2013), https://www.usni.org/magazines/naval-history-magazine/2013/september/historic-aircraft-pioneering-pioneer.

Then there's DARPA, the Defense Advanced Research Projects Agency, a branch of the US Department of Defense established in 1958 after the Soviet launch of Sputnik. DARPA's mission is to keep the US technologically ahead in defense, but its innovations have transformed civilian technology.[180] DARPA created the ARPANET, which became the foundation for the internet. GPS, developed for navy navigation, now guides our everyday commutes. Touch screens, funded by DARPA research, are in our pockets and on our tablets, while their work on voice recognition fuels today's smart assistants.[181]

The agency's contributions continue: Stealth technology has influenced aeronautics, early drone development paved the way for commercial drones, and brain–computer interface research has led to advanced prosthetics. DARPA's investments in satellite technology not only enhanced military communication but also revolutionized weather forecasting and earth observation.

Through these numerous technological innovations, DARPA has not only met its objectives of maintaining military advantage and safeguarding against technological surprises but has also inadvertently driven significant advancements in civilian technology sectors, affecting various aspects of everyday life.

From Alexander's battlefield strategies to DARPA's cutting-edge tech, military innovation reveals a fascinating dual nature, bearing both roses and thorns. These advancements are forged out of a need to defend and conquer—the thorns of conflict driving technological progress. Yet, they often transcend their

180 Paul Szoldra et al., "The Government's Top Scientists Built Some of the Most Amazing Technology We Use Today," Business Insider, January 25, 2016, https://www.businessinsider.com/darpa-creations-2016-1.

181 Glenn R. Fong, "ARPA Does Windows: The Defense Underpinnings of the PC Revolution," in The DARPA Model for Transformative Technologies: Perspectives on the U.S. Defense Advanced Research Projects Agency, ed. William B. Bonvillian, Richard Van Atta, and Patrick Windham (Open Book Publishers, 2019), 146.

original purposes, blooming into roses as they shape civilian life in unexpected ways. Can technology designed for destruction ultimately serve to protect and advance society, or does its very existence ensure that conflict will always evolve alongside it? In this chapter, we examine the paradox of AI-driven military advancements—namely the rise of autonomous drones on the battlefield, the ethical and strategic debates they spark, and the complex concerns surrounding their role in modern warfare.

The Paradox: AI improves battlefield decision-making but raises ethical concerns and risks uncontrollable escalation.

The Roses: Smarter systems save lives and enhance defense capabilities.

The Thorns: Autonomous weapons and decision-making erode accountability and increase the potential for catastrophic errors.

DRONES AND THE NEW FACE OF WAR

Across nearly 2,400 years and thousands of miles, from the fiery catapults of Alexander the Great to the wheat fields of Ukraine, the sounds of war have evolved. Today, a faint hum floats over these fields, emitted by the four propellers of a small, hovering black drone.

This isn't a simple surveillance device but an autonomous, high-tech warrior.

Scanning the ground, the drone runs its calculations with a single question: friend or foe? In mere seconds, sophisticated machine-learning algorithms assess the data, identify a target as hostile, confirm its Russian origin, and execute its deadly mission.

This sleek "black box of death" signals a dramatic new chapter in warfare. Where armies once relied on catapults and long spears, we now see autonomous agents patrolling the skies. It's a shift that reveals how deeply military technology has transformed, bringing autonomous warfare to the front line in ways only imagined before.

A prime example is the deployment of Ukraine's Saker Scout drones, alongside their Russian counterpart, the Lancet. These drones, first confirmed in action in October 2023, marked a major breakthrough—they can conduct strikes independently, without human oversight, a capability that spurred ethical debates worldwide. Though accusations surfaced back in 2020 about similar autonomous systems in Libya, the evidence from Ukraine was the first concrete proof of these technologies in combat.[182]

Initially designed for civilian tasks like crop protection and business automation, the Saker Scout has pivoted since Russia's invasion to meet military needs. These quadcopters, which can carry up to three kilos of explosives and cover a twelve-kilometer range, now operate on a different battlefield. Armed with advanced AI, the Saker Scout learns as it goes, constantly refining its recognition skills to identify a growing array of military targets. In Ukraine's radio-jammed war zones, it can identify and lock onto sixty-four specific Russian military assets, thanks to its machine-learning capabilities.[183]

182 David Hambling, "Ukrainian AI Attack Drones May Be Killing Without Human Oversight," *New Scientist*, October 13, 2023, https://www.newscientist.com/article/2397389-ukrainian-ai-attack-drones-may-be-killing-without-human-oversight/; Maria Cramer, "A.I. Drone May Have Acted on Its Own in Attacking Fighters, U.N. Says," *New York Times*, last updated June 4, 2021, https://www.nytimes.com/2021/06/03/world/africa/libya-drone.html.

183 David Hambling, "Ukraine's AI Drones Seek And Attack Russian Forces Without Human Oversight," *Forbes*, October 17, 2023, https://www.forbes.com/sites/davidhambling/2023/10/17/ukraines-ai-drones-seek-and-attack-russian-forces-without-human-oversight/.

Integrated with Ukraine's Delta intelligence system, the Saker Scout doesn't just navigate and attack autonomously; it also relays crucial battlefield data, vastly reducing the need for human reconnaissance. This tech innovation streamlines the "kill chain" process, accelerating everything from target identification to strike—a speed no human could match.

But the drone's most controversial feature is its capacity to launch attacks without direct human input—a function used selectively, often only when radio interference blocks operator control. This powerful leap in military technology offers strategic advantages, potentially saving countless lives.

Yet it also raises profound ethical questions. Amid the strategic gains, the international community, led by activists and think tanks, is wrestling with the moral implications of autonomous "killer robots." While there's a push for regulation through the UN, the debate is far from settled.

THE REALITY CHECK ON AI-DRIVEN WARFARE

But did the Saker Scout drone live up to all the hype?

In short, no.

The Saker Scout drone was supposed to be the poster child for AI-driven warfare, but it hasn't delivered on those bold promises—at least, not yet. The narrative around AI's potential to reshape the battlefield, particularly in the Russo–Ukrainian conflict, has been a story of grand expectations and stubborn setbacks. What was pitched as a breakthrough in autonomous military operations has turned out to be a lot more complicated—and less effective.

Scout isn't alone in falling short; its Russian counterpart, the Lancet, has faced similar struggles. Both drones were touted as game changers, ushering in a new era of autonomous warfare. Yet, as they hit the field, they revealed critical flaws and limitations that dulled their supposed edge.[184] When it came to the technology, these drones aimed to reduce human input dramatically, with AI making independent calls on target identification and engagement. In practice, though, they've been unreliable. For example, the Lancet's highly promoted "target lock" feature has either failed to work in the field or has been deliberately deactivated—likely due to accuracy issues, as seen in field footage.

The Saker Scout's autonomous targeting capabilities have also struggled to convince, with real-world combat tests falling short of expectations. Many experts now question whether these systems are ready for prime time, suspecting that they're better in theory than in practice.

Economic constraints pile on to these technical challenges. Advanced AI technology isn't cheap, and for Ukraine, which relies heavily on outside funding, the costs make large-scale AI deployment a hard sell. Instead, Ukraine has leaned toward simpler, human-operated drones that, while less advanced, are cheaper and easier to field in higher numbers.

Russia, although somewhat better resourced, faces its own hurdles with budget constraints worsened by sanctions and the long grind of war. The risk of investing heavily in technology that may underperform looms large, especially in a drawn-out conflict where every ruble counts.

For now, the grand vision of AI-driven autonomous warfare remains largely unrealized. It's a reality check that reminds us how often, especially with emerging tech, the hype often outpaces reality.

---

184 Sydney J. Freedberg Jr., "The Revolution That Wasn't: How AI Drones Have Fizzled in Ukraine (So Far)," Breaking Defense, February 20, 2024, https://breakingdefense.com/2024/02/the-revolution-that-wasnt-how-ai-drones-have-fizzled-in-ukraine-so-far/.

CROSSING THE RUBICON

As autonomous weapons edge closer to the battlefield, one question looms large: Has an ethical line already been crossed?

Historically, discussions around autonomous weaponry have focused on the dangers of misuse by malicious forces. But the moral dilemma becomes far more complicated when these technologies are deployed in high-stakes defensive situations—like Ukraine's struggle to protect its sovereignty against a powerful invader. For many Americans, who view Ukraine as the underdog standing up to a larger aggressor, there's a natural empathy for the nation's need for decisive, effective defenses—even if it means bending the ethical boundaries. A recent Pew Research Center study reflects this sentiment, showing that nearly two-thirds of Americans hold a positive view of Ukraine amid the ongoing conflict.[185]

Much like the controversial use of cluster bombs in previous wars, the immediate threat to Ukraine's survival has accelerated the acceptance of autonomous weaponry, further blurring ethical lines.[186] Despite the early setbacks faced by these drones, their potential is hard to ignore. It may be that Ukraine has crossed a kind of ethical Rubicon, a threshold after which these technologies are more likely to be refined than abandoned. Now it's simply a matter of addressing the usual early-stage hurdles—technical glitches, high costs—that accompany any major innovation. Harvard Business School's Clayton Christensen observed that of the 30,000 new products launched annually, 95

185 Jacob Poushter et al., "Americans Hold Positive Feelings Toward NATO and Ukraine, See Russia as an Enemy," Pew Research Center, May 10, 2023, https://www.pewresearch.org/global/2023/05/10/americans-hold-positive-feelings-toward-nato-and-ukraine-see-russia-as-an-enemy/.

186 Hambling, "Ukraine's AI Drones Seek And Attack."

percent face initial failures. Yet, if a product's benefits are clear, ingenuity tends to overcome those early barriers.[187]

As the benefits of autonomous tech in combat become more evident, overcoming these challenges may be just a matter of time. The integration of these systems is poised to grow, potentially transforming warfare itself. But this shift raises urgent ethical questions. If Ukraine sets a precedent, we may see these technologies adopted globally, challenging traditional norms in warfare and sparking debates that are only beginning.

POTENTIAL IN WELL-FUNDED MILITARY PROGRAMS

Despite the initial struggles with autonomous drones in Ukraine, the potential for AI in military applications is still immense—especially for countries with deep technological and financial resources. Nations like the United States and China, which are pouring billions into military innovation as part of a broader tech-driven arms race, are likely to see the most benefit from AI advancements in warfare.

For instance, on March 9, 2023, the Biden-Harris Administration proposed a 2024 budget for the Department of Defense at a whopping $842 billion—$26 billion more than 2023 and $100 billion above 2022. Of that, $145 billion is earmarked for research, development, test, and evaluation (RDT&E), up 4 percent from 2023.[188] This increase underscores the US com-

187 Clara Pilot, "Why 95% of New Products Fail (and How to Prevent It from Happening to You)," *La Razón*, December 13, 2021, https://www.larazon.es/educacion/20211213/qez3zu3nyfgaborsrn44ngvvey.html.

188 Lloyd J. Austin III, "Department of Defense Releases the President's Fiscal Year 2024 Defense Budget," news release, United States Department of Defense, March 13, 2023, https://www.defense.gov/news/releases/release/article/3326875/department-of-defense-releases-the-presidents-fiscal-year-2024-defense-budget/.

mitment to developing cutting-edge technology, materials, and software.

Understanding China's defense spending—especially in technology—remains challenging due to a lack of transparency. Unlike many nations, China does not follow a globally accepted standard for reporting military expenses. While mechanisms like the UN Report on Military Expenditures promote transparency, participation is voluntary. China joined this UN initiative in 2007 but has remained less open about its defense spending than many other countries.[189]

China's defense budget, while more opaque, also reflects significant investment in military technology. In March 2025, the Chinese government reported a defense budget of 1.78 trillion RMB (around $246.5 billion USD), a 7.2 percent increase from the previous year. However, estimates suggest that China's actual military spending may be even higher. For example, the Stockholm International Peace Research Institute pegged China's 2023 defense spending at around $309 billion, while the International Institute for Strategic Studies estimated it at $319 billion—both well above the official figure.[190]

With substantial defense budgets, both China and the US are better positioned to overcome the financial and technological barriers facing autonomous drones, as highlighted by initial challenges in Ukraine. Their considerable resources could speed up the development and refinement of these technologies, making AI-driven and autonomous military systems more capable and reliable. The potential for development suggests

---

189 China Power Project, "What Does China Really Spend on its Military?," Center for Strategic and International Studies, accessed March 20, 2025, https://chinapower.csis.org/military-spending/.

190 China Power Project, "What Does China Really Spend?"

that, with the right focus and resources, today's setbacks could evolve into tomorrow's operational successes.

Since the 1980s, militaries have been deploying partially autonomous weapons, primarily for defensive uses. Today, more than thirty countries employ air and missile defense systems with autonomous functions, built to detect and intercept incoming threats such as rockets, artillery, mortars, missiles, and aircraft. These systems operate autonomously but remain under human oversight, allowing intervention if necessary.

Meanwhile, several countries—including China, France, India, Russia, the United Kingdom, and the United States—are developing stealth combat drones that could potentially operate without human input in future conflicts, targeting air defenses or mobile missile launchers independently. Although autonomous technology for ground vehicles has advanced more slowly than in aerial and maritime applications, future battlefields could feature autonomous ground robots or stationary weapon systems.

The evolution of military technology is likely to continue accelerating. We may soon see autonomous drone swarms coordinating in real time, adapting to shifting battlefield conditions at speeds far beyond human reaction. This shift could push the pace of military operations to levels that surpass human decision-making abilities, potentially reducing the role of human oversight in critical moments.

As warfare becomes more autonomous, new ethical frameworks will be essential to manage these powerful technologies responsibly, shaping the future of conflict and strategy.

THE DEBATE: AUTONOMOUS WEAPONS

The journey of autonomous weapon systems from theory to the battlefield has ignited a passionate debate among military strategists, roboticists, and ethicists. These systems, capable of performing complex tasks like targeting and engaging with minimal human input, are lauded by some military experts for their strategic and tactical benefits, as well as for their potential to save lives. On the other hand, the opposition raises serious ethical and legal concerns. Critics also worry about the potential for malfunctions or hacking, which could lead to devastating unintended consequences. Legally, there are concerns that such weapons could violate international humanitarian law, which prioritizes the protection of human life and dignity.

The debate continues, with each side exploring the far-reaching implications of deploying these advanced technologies in warfare. Let's explore the roses and thorns of autonomous weapons. Through examining these opposing perspectives, we can better understand the complex role autonomous weapons may play in the future of warfare and the importance of guiding their development with strong ethical and regulatory safeguards.

THE ROSES
MILITARY ADVANTAGES

Autonomous weapons bring some serious military advantages to the table, starting with their role as a powerful "force multiplier." With these machines in action, fewer soldiers are needed for a given mission, and each one can accomplish much more. Replacing irreplaceable frontline soldiers with machines could drastically reduce the loss of young American lives, along with the heavy human and financial costs of injuries. Autonomous weapons have the potential to operate with precision and endurance that surpass human capabilities. By eliminating human emotions from the battlefield, these systems might also reduce the risk of war crimes or errors caused by stress.[191]

Not only can autonomous systems operate in tough-to-reach areas, but they also keep human personnel out of high-risk zones, cutting down on casualties.[192] You'll often find these systems tackling what the military calls the "dull, dirty, or dangerous" missions. Think long patrols, handling hazardous materials, or defusing hidden explosives like IEDs—jobs that are critical but can be especially risky or grueling. As US Army Maj. Jeffrey S. Thurnher points out, these lethal autonomous robots have a unique edge: They can operate at speeds humans can't match and can continue to strike even if communications are cut off.[193]

And then there's the cost-effectiveness, where the numbers are significant. As David Francis wrote in the *Fiscal Times*, each soldier in Afghanistan costs the Pentagon roughly $850,000 per year. Meanwhile, a TALON robot—a small weaponized rover—comes in at $230,000.[194] At one quarter the cost, robots are replaceable, and they don't need time off or healthcare—nor do they have wives, husbands, children, friends, or family.

The Department of Defense is also investing in advancing the autonomy of these systems. They're focusing on key areas like perception, planning, learning, human–robot interaction,

---

191 Amitai Etzioni and Oren Etzioni, "Pros and Cons of Autonomous Weapons Systems," *Military Review* 97, no. 3 (May–June 2017): 72–81, https://www.armyupress.army.mil/Portals/7/military-review/Archives/English/pros-and-cons-of-autonomous-weapons-systems.pdf.

192 Etzioni and Etzioni, "Pros and Cons," 72.

193 Jeffrey S. Thurnher, "No One at the Controls: Legal Implications of Fully Autonomous Targeting," *Joint Force Quarterly* 67, no. 4 (2012): 83, https://ageoftransformation.org/content/files/Portals/68/Documents/jfq/jfq-67/jfq-67_77-84_thurnher.pdf.

194 Etzioni and Etzioni, "Pros and Cons," 72.

and coordination between multiple agents.[195] These improvements are expected to make these systems even more effective and efficient.

For aerial missions, autonomous systems hold even more potential. As Air Force Maj. Jason S. DeSon explained in the *Air Force Law Review*, robot pilots have none of the limitations of human ones. Fatigue, physical strain from high-G maneuvers, and the intense focus required for aerial combat are no problem for robots.[196] In fact, autonomous aircraft can perform random, unpredictable maneuvers, potentially throwing off enemy pilots who rely on anticipating human patterns. This ability to go beyond human physical limits and predictability makes autonomous systems a strategic asset in combat.[197]

MORAL JUSTIFICATIONS

Some military experts argue that autonomous weapons might actually act more ethically than human soldiers on the battlefield. One reason is that robots don't have a survival instinct, which can reduce the urge for preemptive strikes based on fear or self-preservation. Roboticist Ronald C. Arkin believes that autonomous robots in the future could behave more ethically because they lack that "shoot first, ask questions later" mindset that fear can trigger in humans.

Without emotions clouding their judgment, autonomous

---

195 Defense Science Board, *Task Force Report: The Role of Autonomy in DoD Systems* (Defense Science Board, July 2012), 31–54, https://apps.dtic.mil/sti/tr/pdf/ADA566864.pdf.

196 Jason S. DeSon, "Automating the Right Stuff? The Hidden Ramifications of Ensuring Autonomous Aerial Weapon Systems Comply with International Humanitarian Law," *Air Force Law Review* 72 (2015): 85–122, https://www.afjag.af.mil/Portals/77/documents/Law%20Review/AFD-150721-006.pdf?ver=ilF66sB6p6edwlF52Txx6g%3d%3d.

197 Etzioni and Etzioni, "Pros and Cons," 73.

systems could process vast amounts of information more accurately and make decisions based on logic alone. US Army Lt. Col. Douglas A. Pryer suggests there's a real ethical upside to using robots in high-stress combat zones. Neuroscience research has shown that extreme stress can impact the brain's ability to control impulses, sometimes leading soldiers to actions like aggression or misconduct they'd typically avoid under normal conditions. Robots, on the other hand, wouldn't experience this stress, allowing them to operate under consistent ethical guidelines.[198]

Moreover, robots might be more reliable in reporting any ethical violations, potentially boosting accountability in combat. This commitment to rules without emotional interference could make autonomous weapons not just effective but ethically preferable on the battlefield.

While the potential benefits of autonomous weapons are significant, they've sparked equally serious concerns across the globe.

THE THORNS
ETHICAL AND LEGAL CONCERNS

As autonomous weapons become more prevalent in military operations, they're igniting ethical and legal debates, particularly around international humanitarian law. Principles like the Principle of Distinction—which requires differentiating between combatants and civilians—are central to these concerns, as is the issue of accountability. Experts like computer

---

198 Douglas A. Pryer, "The Rise of the Machines: Why Increasingly 'Perfect' Weapons Help Perpetuate Our Wars and Endanger Our Nation," *Military Review* 93, no. 2 (March–April 2013): 15, https://www.armyupress.army.mil/Portals/7/military-review/Archives/English/MilitaryReview_20130430_art005.pdf.

scientist Noel Sharkey and ethicist Robert Sparrow worry that autonomous systems might blur the lines of responsibility in wartime, potentially leading to civilian casualties with no clear accountability.[199]

This debate reflects a broader dilemma: While autonomous systems offer strategic advantages, they also raise ethical red flags, especially when human involvement in life-or-death decisions is minimized. As technologies like autonomous drones advance, ensuring they operate within strict ethical boundaries becomes crucial. These systems need transparent decision-making processes to prevent the reinforcement of biases that could have catastrophic consequences on the battlefield.

This shift toward autonomous military technology demands a balanced approach, one that boosts operational efficiency but also upholds stringent ethical and legal standards. Accountability remains key—especially when unintended casualties are involved—and developing AI that serves both tactical goals and moral principles is essential to creating a just and responsible framework for the future of warfare.

LEGAL FRAMEWORKS

With AI technologies advancing rapidly in the military sphere, existing legal frameworks are struggling to keep up.

Questions around intellectual property for AI-generated content, liability for damages caused by AI decisions, and the use of personal data in AI training all push the boundaries of current laws. The European Union has taken notable steps

---

199 Noel Sharkey, "Saying 'No!' to Lethal Autonomous Targeting," *Journal of Military Ethics* 9, no. 4 (2010): 369–383, https://doi.org/10.1080/15027570.2010.537903; Robert Sparrow, "Killer Robots," *Journal of Applied Philosophy* 24, no. 1 (February 2007): 62–77, https://doi.org/10.1111/j.1468-5930.2007.00346.x.

with the General Data Protection Regulation (GDPR)—widely regarded as the world's strongest privacy law—which addresses automated decision-making and profiling.

Yet, despite these efforts, major gaps remain.

Key issues like cross-border law enforcement, accountability for autonomous systems, and even the concept of AI person-hood—the idea of granting AI legal rights and responsibilities akin to those of humans or corporations—are still uncharted territory.[200]

Addressing these challenges requires a coordinated international approach. We need comprehensive legal frameworks that can keep pace with AI's rapid evolution while safeguarding human rights and societal values. These frameworks should outline clear standards for AI development, establish liability laws for when AI systems malfunction, and introduce ethical guidelines to ensure AI innovations benefit humanity as a whole.

THE GLOBAL DIALOGUE ON AUTONOMOUS WARFARE

Around the world, NGOs and international organizations are stepping up to push back against autonomous weapons, rallying global support to keep these systems in check as they recognize the ethical Rubicon we've already crossed and the urgent need to address the growing risks before they spiral further out of control.

Leading the charge since 2013 is the Campaign to Stop Killer Robots, a coalition of NGOs working to secure a preemptive

---

200 Simon Chesterman, "Artificial Intelligence and the Limits of Legal Personality," *International and Comparative Law Quarterly* 69, no. 4 (October 2020): 819–844, https://doi.org/10.1017/S0020589320000366.

ban on lethal autonomous weapons.[201] Alongside this campaign, the United Nations has been facilitating discussions through the Convention on Certain Conventional Weapons, hosting forums to explore potential regulations for these powerful technologies.[202]

Public opinion is a critical force in this debate, especially in regions like Europe, where strong public resistance has fueled calls for tighter legislative controls. In response to these concerns, the European Parliament has even debated stricter regulations for autonomous weapons.[203]

Influential voices have also added weight to the movement. In July 2015, an open letter sent shock waves through the AI community, calling for a ban on autonomous weapons. This wasn't just any petition—it was backed by influential voices like Elon Musk of Tesla, Apple co-founder Steve Wozniak, celebrated physicist Stephen Hawking, and MIT's Noam Chomsky.[204]

Alongside these figures, more than three thousand AI and robotics researchers signed on, warning that the rapid advances in AI could soon make fully autonomous weapons a reality. They described these systems as a potential "third revolution" in warfare, following gunpowder and nuclear weapons, and

---

201 Mary Wareham, "Stopping Killer Robots: Country Positions on Banning Fully Autonomous Weapons and Retaining Human Control," Human Rights Watch, August 10, 2020, https://www.hrw.org/report/2020/08/10/stopping-killer-robots/country-positions-banning-fully-autonomous-weapons-and?gad_source=1&gclid=CjwKCAjw-qi_BhBxEiwAkxvbkFZGxL2oIk_cg8umTZz9RlITHTjz_qjJMhn-5uEqpRChqwB-vqnsiRoCZ6YQAvD_BwE.

202 "Autonomous Weapons Are a Game-Changer," The Economist, January 25, 2018, https://www.economist.com/special-report/2018/01/25/autonomous-weapons-are-a-game-changer.

203 European Parliament, "Guidelines for Military and Non-Military Use of Artificial Intelligence," press release, January 20, 2021, https://www.europarl.europa.eu/news/en/press-room/20210114IPR95627/guidelines-for-military-and-non-military-use-of-artificial-intelligence.

204 "Autonomous Weapons: An Open Letter from AI and Robotics Researchers," Future Timeline (blog), July 28, 2015, https://futuretimeline.net/blog/2015/07/28.htm.

emphasized the serious risks involved.[205] Their main concern? That an AI arms race could tarnish AI's reputation and limit its potential for positive, peaceful uses. The letter specifically pushed for a ban on offensive autonomous weapons that lack meaningful human oversight. This powerful message amplified public awareness and put additional pressure on lawmakers to act.

Two years earlier, in April 2013, the United Nations issued a report urging countries to consider moratoriums on lethal autonomous robots until proper international guidelines could be established.[206] Later that year, scientists from thirty-seven countries published the "Scientists' Call to Ban Autonomous Lethal Robots," spotlighting the limitations of current technology. They stressed that robots struggle to distinguish combatants from civilians and can't reliably make proportional decisions about force. The declaration ended on a strong note, insisting that "decisions about the application of violent force must not be delegated to machines."[207]

As these technologies continue to advance, the urgency to establish regulatory frameworks only grows. Although a complete ban on autonomous weapons may not be feasible, practical measures are essential to mitigate their most severe risks. Without these measures in place, we risk seeing conflicts spiral out of human control. The use of autonomous drones, like Ukraine's Saker Scout, highlights the need for swift, coordinated action from the international community. However, despite years of discussion, a global consensus on how to manage the risks of

205 Future Timeline (blog), "Autonomous Weapons: An Open Letter."

206 Christof Heyns, "Report of the Special Rapporteur on Extrajudicial, Summary or Arbitrary Executions," United Nations Human Rights Council, April 9, 2013, https://www.ohchr.org/sites/default/files/Documents/HRBodies/HRCouncil/RegularSession/Session23/A-HRC-23-47_en.pdf.

207 Etzioni and Etzioni, "Pros and Cons," 75.

autonomous weapons remains elusive. This is particularly concerning given the technological rivalry between the US and China, which could easily spark an AI arms race.

This ongoing dialogue is crucial to prevent a future dominated by machine-driven warfare and to implement meaningful controls that curb an unchecked arms race among global powers.

FUTURE POSSIBILITIES: FROM BATTLEFIELD TO BREAKTHROUGHS

Despite the global dialogue on autonomous weapons highlighting their risks, we must also ask whether these technologies can be repurposed to serve global needs in peacekeeping, disaster relief, environmental sustainability, and more. If managed wisely, what now serves the battlefield could evolve into a powerful force for humanitarian goals.[208]

Today's arms race in autonomous systems and AI, especially between superpowers like the US and China, primarily aims at military enhancement. Yet it also holds the potential to spark critical breakthroughs that address major global issues, from climate change to healthcare. But to unlock this potential, a focus on ethics and strong international regulations is vital. These safeguards could prevent military escalation and help ensure that these technologies serve human welfare.

The benefits of this technological race are broad and far-reaching. Advances in military robotics could reshape manufacturing, healthcare, and disaster response, where precision and reliability are critical. Combat-driven AI advancements

---

208 Nermin Hajdarbegovic, "Cold War Tech: It's Still Here, and Still Being Used," *Toptal* (blog), accessed March 20, 2025, https://www.toptal.com/cybersecurity/cold-war-tech-cyberwarfare-cybercrime.

might improve environmental monitoring and wildlife conservation. Similarly, the cybersecurity leaps needed for cyber warfare could offer vital protections for public and private sectors alike. Military surveillance tools, adapted for environmental monitoring, could aid sustainable development by tracking resource use and ecosystem health.

The precision demanded by military robotics could lead to medical robots and telemedicine solutions, especially valuable in remote or underserved areas. The drive for efficient energy in military contexts may also spur innovations in renewable energy and storage, crucial for reducing fossil fuel reliance. Finally, secure communication systems designed for military operations could strengthen civilian networks, boosting resilience in emergencies and disaster responses.

These examples highlight how military-driven technologies can have dual uses, benefiting both defense and civilian sectors. With strategic redirection, military innovation can lead to breakthroughs that uplift society, turning tools of conflict into assets for global progress.

CLOSING THOUGHTS: WHEN WAR-DRIVEN INNOVATIONS WORK FOR US ALL

Throughout history, warfare has driven remarkable technological advancements, leaving legacies that extend far beyond the battlefield—from Alexander the Great's tactical innovations to the strategic breakthroughs of the Cold War.

Today, the global race in autonomous systems and AI, particularly between powers like the US and China, continues this trend. With wise guidance, this competition holds the potential to spark innovations that not only serve military purposes but also benefit society in areas like healthcare, environmental sustainability, and public safety. The crucial task is to direct these advancements toward civilian uses, while updating ethical standards and international regulations to match the rapid pace of technological progress.

Reflecting on the dual-use nature of past innovations reminds us that, with careful management, the jagged edge of military technology can lead to meaningful improvements in human welfare. This approach is essential—not only to mitigate the risks of escalation but also to ensure these powerful technologies contribute positively to global progress and connectivity.

Across the paradoxes presented in Part Two, one thing becomes clear: AI is not inherently good or bad—it is what we make of it. As we navigate these complex trade-offs, the question is not just how to solve these paradoxes but how to do so in a way that reflects our values and shapes a future we can be proud of.

CONCLUSION

In 1633, Galileo Galilei found himself on trial, accused of heresy for his revolutionary ideas about the universe. Under a sweltering June sky, the courtroom filled with murmurs and jeers, while commoners clutched their crosses, hoping to fend off what they saw as his "evil" influence. When the judge finally pronounced his sentence, Galileo's response was barely a whisper: *"E pur si muove"—"And yet it moves."*

His quiet defiance, confirming that Earth orbits the sun, was a truth too weighty for his accusers to grasp. Galileo was branded a threat—a prophet of ideas far ahead of his time, yet treated as a pariah for challenging the comfort of convention.

PROPHETS AND PARIAHS OF PROGRESS

Galileo's story is far from unique. History is rich with figures who dared to see beyond the accepted truths of their day—Socrates, Jesus, Nikola Tesla, Charles Darwin, Alan Turing, Martin Luther King Jr., to name a few. Each one challenged established norms. And each one faced persecution, imprisonment, or, in some cases, death.

Socrates's radical approach to questioning laid the foundations of Western philosophy, yet he was accused of corrupting Athenian youth and ultimately executed. Tesla's groundbreaking work on alternating current clashed with the dominant financial interests of the time, sidelining him despite his revolutionary contributions. Darwin, proposing that species evolved through natural selection, was scorned for challenging creationist beliefs. And Turing, a pioneer of computer science, faced persecution due to his sexuality, despite his work in breaking the Enigma code and shaping the future of computing. Even Martin Luther King Jr., who inspired millions with his vision of equality, met with fierce resistance and, ultimately, assassination.

They were prophets—bearers of a future that society wasn't ready to accept.

Yet they were treated as pariahs—dismissed and punished for their ideas.

These figures were seen as dangerous because they questioned established truths. In time, however, their ideas reshaped entire fields, proving that the very notions society initially feared were, in fact, tipping points that propelled humanity forward. Malcolm Gladwell's "tipping point" theory comes to life in their stories, illustrating how pivotal outliers often set the stage for profound societal shifts.

THE PARADOX OF PROGRESS

As we navigate these historical accounts, a pattern emerges—the paradox of progress. For every step forward, there is a pushback; for every groundbreaking idea, there are unforeseen consequences.

These tipping points remind us that while progress drives humanity to greater heights, it often comes at a cost. Each prophet who reshaped society, from Galileo to King Jr., challenged the collective mindset, and with their progress came both roses and thorns: admiration for their contributions but resistance and suffering along the way.

This paradox leads us to our current age, where artificial intelligence and autonomous technologies are emerging as the latest tipping points. Like the visionaries of the past, AI holds the potential to change society fundamentally. Yet, it faces the same scrutiny and fear once reserved for prophets and pariahs, raising the question: Will we see AI as a tool for progress, or will we treat it as a threat?

In other words, will we view AI as a prophet or a pariah?

AI, with its capacity to analyze, predict, and even mimic human thought, is both revered for its potential and feared for its unknowns. Like Galileo's telescope or Darwin's theories, AI challenges our current understanding and stirs discomfort by asking us to redefine our boundaries. In fields like healthcare, environmental monitoring, and education, AI has shown it can create profound improvements, tackling issues faster and with greater precision than ever before. Yet, like all disruptive innovations, AI raises thorny questions—about privacy, ethics, and our relationship with technology.

Imagine AI as a modern-day prophet, with the power to anticipate and help us avoid future mistakes—much like it could have helped steer away from the dark legacy of slavery in the American colonies or the unforeseen consequences of past technologies. Could AI's predictive abilities help us make wiser choices and sidestep the pitfalls we've seen throughout history?

Or, like prophets past, will AI be shunned as a disruptive force, feared rather than embraced?

Today, AI stands at the same crossroads as Galileo, Tesla, and Turing. It has the potential to be a prophet of transformative progress or a pariah, feared for its thorns and cast aside despite its roses.

As we push forward in developing AI for everything from peacekeeping to environmental protection, we must grapple with these dualities and ask ourselves, Are we ready to listen to what AI has to tell us?

IT'S UP TO *US*

The experiences of past visionaries serve as a reminder to approach AI with both openness and caution. In the end, AI embodies the same paradox faced by every groundbreaking idea in history. Its potential to unify or divide, to benefit or harm, depends on the path *we* choose. It's up to *us* to determine whether this tipping point leads to growth or division.

As we stand at this tipping point, the future rests on how we choose to wield AI's power. By recognizing that every leap forward has both light and shadow, we can strive to make informed choices that steer this innovation toward a positive impact on humanity.

And so, we are left with a choice much like those societies faced with their own prophets and pariahs. Will we see AI as a prophet of progress, with the power to uplift and guide, or a pariah to be mistrusted and marginalized? Like a rose with its thorns, AI's potential comes with both risks and rewards.

Our choices today will shape not only the trajectory of AI but the destiny of humanity itself. We must decide: Will AI be a rose or a thorn, a force for unity or a divider?

The future lies in our hands—and with it comes a responsibility.

As we turn the final page of this exploration into the dual nature of artificial intelligence, let's not see it as the end, but as the beginning of an essential dialogue. *The Paradox of Progress: The Roses and Thorns of Artificial Intelligence* was never meant to provide all the answers—it was written to spark questions and ignite discussions that we need to have in our communities, workplaces, and educational institutions. It's an invitation to carry this conversation forward—to coffee shops, living rooms, offices, and classrooms around the world.

By weaving these discussions into the fabric of our daily lives, we can collectively navigate the complexities of AI. Together, we can ensure that our technological advancements align with our most cherished human values, ethically shaping a future where AI enriches rather than diminishes the human experience. My hope is that this book serves as a catalyst for thoughtful deliberation, inspiring us to harness the potential of artificial intelligence responsibly and steer societal progress toward outcomes that benefit everyone.

"Technology is nothing. What's important is that you have faith in people, that they're basically good and smart, and if you give them tools, they'll do wonderful things with them."

—STEVE JOBS

ACKNOWLEDGMENTS

Writing *The Paradox of Progress* has been a journey filled with introspection and support from a remarkable circle of family, colleagues, and professionals. This book reflects not only my thoughts but also the encouragement and wisdom I've received along the way.

First and foremost, I extend my deepest gratitude to my family, the core team behind everything I do. My wife, Kim Escudero, MD, has been more than a spouse; she has been my rock for more than thirty years. Always working diligently behind the scenes, Kim's support and companionship have made all my pursuits possible, and this book is no exception.

To Mia, my eldest daughter, your boundless positivity has often been the boost I needed. Your simple, "Hi, Dad, how was your day?" recharges my spirit, reminding me of the power of small gestures. Your constant cheer and resilience bring light into our lives, inspiring me daily and proving how everyday kindness can be profoundly impactful.

Dane, your commitment to competitive cross-country skiing is a testament to the values of persistence and dedication—core principles of innovation. You inspire me daily to push my limits to redline, a lesson I carry into every chapter I write. I eagerly anticipate the world discovering what I have long known: You are a force set to ignite and transform whatever path you choose.

Olivia, watching you is like seeing a reflection with an intensity that astonishes me. Your work ethic and that memorable episode in the grocery store on Thanksgiving 2021 sparked the initial idea for this book. Your honest critiques of my all-too-human traits propelled me to start writing, and for that, I am profoundly thankful. Your drive and dedication are mirrored in the way you approach every challenge—head-on and with unmatched commitment. I truly wonder what you will create in this world.

My journey through life and learning has been profoundly shaped by my educators, who taught me not just to think, but to challenge and broaden my horizons. I owe a debt of gratitude to my parents, Peter and Lillian Karch, whose wisdom and guidance lit my path from the beginning. The teachings of Bill Riedel; Jack Delahay, MD; Brian Evans, MD; and the entire Georgetown University Department of Orthopedics have left a deep mark on my professional life. I remember the late Jeff Mast, MD, a master surgeon whose skills and dedication to invention and innovation continue to inspire me. Jim Lubowitz, MD, and Dan Guttman, MD, also deserve special mention for their mentorship and knowledge.

As a middle-aged student returning to the classroom after a long hiatus, my academic growth was further nurtured by my professors at the Harvard Business Analytics Program—Marco Iansiti, PhD; Karim Lakhani, PhD; Michael Tushman, PhD; Michael Parzen, PhD; and Tsedal Neeley, PhD—who opened my eyes to the potential of artificial intelligence. I am forever grateful to my teachers and mentors, who instilled in me the values of curiosity and lifelong learning. Your lessons extend beyond classrooms and textbooks, echoing through every challenge I face.

I believe in the untapped power of teams, where individuals use their talents to multiply their collective effect. As the wax technician for the Tahoe Endurance Nordic cross-country ski team, I've had the pleasure of applying scientific principles to enhance performance, re-creating the innovation cycle of success, failure, and progress through the process of testing and iteration. The team's spirit and drive are a constant source of inspiration, demonstrating that "Victory by Design" is the perfect blend of passion, science, and teamwork. For each member of this elite team, but especially the coaches Quinn Lehmkuhl, Julian Bordes, and Ettiene Bordes, thank you for reenergizing me each winter weekend, allowing me to return each Monday to start the workweek with newfound strength and resolve.

Teams exist in many forms in my personal and professional life, and recently, I was introduced to yet another high-performance team at Scribe Publishing. Mark Chait, executive editor, you took my manuscript and returned it with insights that were both challenging and enlightening. Emmy Koziak, my publishing manager, your precision and timeliness made this process smooth and efficient.

And then there are special individuals within great teams, ones that excel and propel the whole team forward. This book would not be what it is today without the unique talents of Aleks Mendel, my editor for all things written. Your approach transcends mere red-pen corrections; you engage deeply in the process of mutual learning, making each writing session an enjoyable and enlightening experience. Your guidance has not only sharpened my writing but has also made me a better storyteller.

There are also team members I've never met, yet their

contributions laid the groundwork for a book like this to be written. To the unsung heroes of technology—computer scientists, mathematicians, technicians, and the countless unnamed coders—your creativity and dedication shape the digital world we live in. Your efforts remind us that technology is ever-evolving, driven by a continuum of innovation and resilience from those who came before us.

I am also grateful to my team at SteriTools—Leigh Biedeman Moss, Chris Wiley, Negin Bemanzadeh, Charles Brooks, Leslie Morales, and my close friend Lon White. Your collective efforts behind the scenes have been crucial in balancing this project with my other commitments. The engineers at NorthSky Design in the UK deserve a special mention for turning ideas into tangible innovations. Watching the entire SteriTool team work, fail, iterate, fail again, reiterate, and ultimately succeed is like witnessing the cycle of innovation in real time every week. Your relentless process of trial and refinement has been a fundamental inspiration for this book.

Brooke Thompson and the Velys Depuy robotics and computer navigation team at Golden State Orthopedics, thank you for introducing me to the fascinating world of robotics in surgery. Bill Salmon, your dedication to ensuring our equipment is ready despite formidable challenges is deeply appreciated. Bill is like a postman—through rain, sleet, snow, and treacherous mountain road conditions, he always gets there to turn the robot on. Bill is a machine, and because of the man, the machine can work.

In *The Paradox of Progress*, we explore the critical balance between advancing technology and maintaining the irreplaceable human touch. This theme is vividly exemplified within my orthopedic practice and surgical team, where each member plays a vital role in blending these elements.

In the daily operations of my clinical practice, Christine MacDonald, PA-C, and Karly Dawson, PA-C, exemplify patient-focused efficiency. They, along with Michael McMahon, AT, our office manager, and clinical assistants Kathleen Hopps; Zack Smitheman, AT; Katie Dease; Sarah Lang, AT; Gage Ramirez, AT; Heather Roberts; Laura Costa; Sandra Guzman; and Ruiko Hosaka, AT, ensure everything runs seamlessly. Tina Alec, RN, the "big gear" of our team, orchestrates these dynamics with relentless dedication, proving how crucial human interaction is for interpreting and responding to the nuances of patient care that technology alone cannot manage.

On the surgical side, the coordination between Dr. Nat Parker, Dr. Larry Silver, and Sandra Bowman, RN, along with other team members, underscores how technology coupled with human expertise can enhance surgical outcomes. While advanced medical technologies provide the tools for precise operations and diagnostics, it is the surgeons, anesthesiologists, and nurses who interpret these outputs and make critical decisions, ensuring that our technological integration translates into tangible patient benefits.

At Mammoth Orthopedic Institute, my colleagues Tim Crall, MD; Brian Gilmer, MD; Stephen Knecht, MD; Tyler Williamson, MD; Rucker Stagger, MD (our current fellow); Richard Brown, MD; and David Hackley, MD, further embody this integration. Their commitment to patient care and team building enhances our collective ability to address health challenges effectively. As we incorporate advanced technologies, their efforts ensure we never overlook the essential human elements—empathy, judgment, and collaboration—that are fundamental to delivering exceptional healthcare.

As we advance with machine intelligence, it's crucial to remember that progress in all sectors, but especially healthcare, is not just about technological enhancement but also about preserving the human touch. This balance is what truly drives forward the cycle of innovation and improvement, ensuring our healthcare system remains not only technologically adept but deeply human-centered.

And finally, to my patients, who face pain and adversity with remarkable resilience—your experiences fuel every chapter of this book. Thank you for reminding me daily of the profound impact that compassionate medical care can have. I am hopeful that, with ethical application, this new wave of technological AI advancements will significantly enhance the health and well-being of us all.

This book is a tribute to all of you, whose stories, support, and insights have been instrumental in shaping this narrative. Thank you for being part of my own paradox of progress. We live in an era where the acceleration of technological and social change presents both monumental opportunities and significant challenges. This "paradox" encapsulates our journey through an age where every advancement seems to pose new dilemmas—where our quest for knowledge and mastery often leads us to unforeseen consequences. As we navigate this complex landscape, your contributions have not only informed this work but also illuminated the path forward. Together, we explore how the very tools we create to bring us together can also divide, how the pursuit of betterment can sometimes lead to complexity rather than simplicity. Thank you for being an integral part in this critical examination of our times, and for helping to forge a narrative that seeks not just to question but also to understand.

ABOUT THE AUTHOR

DR. MICHAEL KARCH is a highly accomplished orthopedic surgeon, specializing in Johnson & Johnson Depuy Velys computer-navigated and robotic hip and knee replacements. In addition to his medical expertise, Dr. Karch is an entrepreneur, inventor, author, and dedicated international disaster responder. He coinvented the SmartDrill and SteriTools technologies, holding numerous patents in the medical device field. Dr. Karch is also a co-founder of the Mammoth Orthopedic Institute and serves as a faculty member in their orthopedic fellowship teaching program. He serves as an adjunct associate professor of orthopedic surgery at Georgetown University School of Medicine.

Dr. Karch's commitment to global health is exemplified through his rapid response to international disasters. He was one of the first physicians on-site at Ground Zero following the September 11, 2001, attacks and has led Forward Surgical Teams to multiple disaster zones worldwide. He co-founded Mammoth Medical Missions, a 501(c)(3) organization that provides emergency medical care in disaster and conflict areas. His leadership has earned him two letters of commendation from the president of the United States and international recognition from the United Nations, as well as from the governments of Iraq, Kurdistan, Nepal, Mexico, and the Philippines.

An accomplished author, this is Dr. Karch's third book. His other books include *Tangible Heroes: The Force Multiplying Power of Ordinary People with Extraordinary Impact* and *Artificial Intelligence for the Everyday Person: Demystifying AI*. He is currently studying artificial intelligence and business analytics at Harvard Business School; at MIT, he is studying machine learning in their executive-level certificate education program.

Outside his professional life, Dr. Karch is an avid endurance athlete, with fifty-seven marathons, six Iron Distance Triathlons, and numerous ultradistance running, cross-country skiing, and mountain biking events to his name. He has completed a solo multiday unsupported 211-mile run on the John Muir Trail in the High Sierra mountains of California as well as an ultra-run along the Inca Trail in the Peruvian Andes. He is also a two-time finisher of the 135-mile Badwater Ultramarathon across Death Valley, often regarded as the world's toughest footrace.

Dr. Karch lives on his family ranch in Mammoth Lakes, California, with his wife, Kim, and their three children, Mia, Dane, and Olivia.

www.ingramcontent.com/pod-product-compliance
Lightning Source LLC
Chambersburg PA
CBHW030502210326
41597CB00013B/759